BROWN'S SIGNALLING

BROWN'S
SIGNALLING

How to Learn the
INTERNATIONAL CODE
OF VISUAL AND SOUND SIGNALS

Based on Information contained in the 1969
International Code of Signals

REVISED BY

J. E. MILLIGAN

B.Sc., Dip.Ed., M.Ed., M.N.I., Master Mariner, Lecturer, *Glasgow College of Nautical Studies*. Previously Senior Lecturer, Hong Kong Polytechnic.

GLASGOW
BROWN, SON & FERGUSON, LTD., NAUTICAL PUBLISHERS
4-10 DARNLEY STREET

Copyright in all countries signatory to the Berne Convention
All rights reserved

First Edition	–	– 1933
Eighth Edition	–	– 1969
Revised Edition	–	– 1974
Ninth Edition	–	– 1979
Revised Edition	–	– 1988

ISBN 0 85174 350 1 (Ninth Edition)

© 1988 BROWN, SON & FERGUSON, LTD., GLASGOW G41 2SD
Made and Printed in Great Britain

CONTENTS

CHAPTER I
Description of Flags—Definitions 9

CHAPTER II
Types of Signals—Signal Letters—Important Two-Letter Groups 13

CHAPTER III
The Use of Numeral Pendants—The Use of Substitutes 16

CHAPTER IV
Alphabetical Signals 21

CHAPTER V
Arrangement of Code Book—Examples of Coding and Decoding 22

CHAPTER VI
Positions for Hoists—How to Call—How to Answer Signals—How to Complete Signal—When Signals are not Understood—Communication by Local Signal Codes—Communication by Flags between Men-of-War and Merchant Vessels—Questions Relating to Flag Signalling—Answers 27

CHAPTER VII
Semaphore Signalling 34

CHAPTER VIII
Morse Signalling—Explanation of Use of Procedure Signals and Signs—Letter- and Figure-Spelling Tables—Signs for Voice-Form of Message—Examples of Transmission of Messages—I.A.N.S./D.Tp. Examination in Signalling 38

CHAPTER IX
Morse Signalling by Hand Flag or Arms 53

CHAPTER X

Radiotelephony 55

CHAPTER XI

Sound Signalling—Questions Relating to Morse Signalling—Answers Relating to Morse Signalling 56

CHAPTER XII

Single Letter Signals—Single-Letter Signals with Complements—Icebreaker Single-Letter Signals—General Code—Table of Complements 59

CHAPTER XIII

Medical Section—Medical Code—Distress Signals 108

CHAPTER XIV

Firing Practice and Exercise Areas 130

CHAPTER XV

Emergency Radiotelegraph and Radiotelephone Procedures 133

CHAPTER XVI

About Flags—Relative to the Origin of the Union Flag in its Present Form—British Naval Signalling Flags—International Salutes—Flags to be Flown by British Merchant Ships—Flag Etiquette 138

CHAPTER XVII

International Regulations for Preventing Collisions at Sea 151

CHAPTER XVIII

General Notices 184

INTRODUCTION

THE 1969 International Code of Signals is intended to cater primarily for situations related essentially to safety of navigation and persons, especially when language difficulties arise. It is suitable for transmissions by ALL means of communication, including Radiotelephony and Radiotelegraphy. The revised code embodies the principle that each signal has a complete meaning.

This code is published in nine different languages, viz. English, French, Italian, German, Japanese, Spanish, Norwegian, Russian and Greek. *Brown's Signalling* contains all the instructions relative to Visual and Sound Signalling and is compiled in such a manner to assist those who desire to acquire proficiency in modern signalling; the subjects are arranged in order best suited for that purpose.

This book is frequently revised to ensure it is up to date in all respects.

NOTE—Certain official matter in this book is taken from The International Code of Signals by permission of the Controller of H.M. Stationery Office London.

NATIONAL COLOURS
PLATE 1

Union Flag

White Ensign

Blue Ensign

Royal Air Force

Red Ensign

Pilot Jack

Argentine

Belgium

Burma

Chile

China

Denmark

France

Greece

Holland or Netherlands

India

Italy

Japan

Norway

Pakistan

Panama

Portugal

Spain

Sweden

United States

U.S.S.R.

INTERNATIONAL CODE OF SIGNALS
PLATE 2

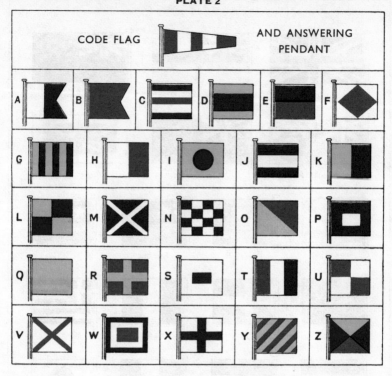

NUMERAL PENDANTS.

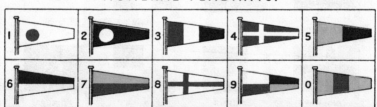

SUBSTITUTES.

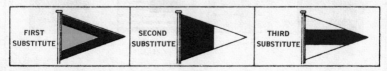

BROWN'S SIGNALLING

CHAPTER I

DESCRIPTION OF FLAGS—DEFINITIONS

Signals are conventional signs or symbols which, when exhibited singly or collectively, convey informations, and the process employed in making the signs is termed signalling. The mediums generally used are flags, shapes, light and sound, the system adopted varying according to circumstances.

In the following pages, visual and sound signalling will be explained, beginning with flags.

Each letter of the alphabet can be represented by a flag which, when hoisted, may be read as the letter it represents, or convey particular information. In the latter case it is necessary to have a Code to which reference can be made to obtain the import of the signal. Obviously where a Code is employed the signal need not be confined to one flag, otherwise the Code would be restricted to 26 messages.

Where two different alphabetical flags are used to form a signal, more signal combinations are available, and still more if these Two-letter groups precede a figure. The only Three-letter groups are in the Medical Section and are easily recognised as they start with the letter M.

This book, while mainly for the use of ships, also provides a number of signals for the use of aircraft and ships.

A complete set of signal flags consists of 26 alphabetical flags, 10 numeral pendants, 3 substitutes, and the answering pendant.

On Plate 2 the flags of the International Code are given in colour. Each of the flags should be memorised so that they may be recognised at a glance.

Before proceeding further, it will be expedient to describe the flags of the Code and define certain terms used in the process of signalling.

ALPHABETICAL FLAGS

A is a burgee, one of the only two burgees in the International Code. It is White and blue, divided vertically. The white half is next to the mast (or at the hoist), the blue half is at the fly which is swallow-tailed.

B like A is a burgee; it is all red.
The remaining flags of the alphabet are rectangular in shape.

C is a flag divided horizontally with blue, white, red, white and blue bars.

D is a yellow flag with a central broad horizontal blue stripe.

E is a flag, blue in the upper half and red in the lower half.

F is a white flag with a red diamond whose corners touch the edges of the flag.

G is a flag, divided vertically with yellow and blue bars, yellow at the hoist, blue at the fly.

H is a white and red flag, divided vertically, the white at the hoist and the red at the fly.

I is a yellow flag with a black circle in the centre.

J is a blue, white and blue flag, divided horizontally (white stripe in centre).

K is a yellow and blue flag, divided vertically, yellow at the hoist, blue at the fly.

L is a yellow and black quartered flag, yellow at the upper hoist, yellow at lower fly.

M is a blue flag with a white St. Andrew's Cross.

N is a blue and white chequered flag. It has sixteen squares, the square in the upper hoist and lower fly being blue.

O is a yellow and red flag, divided diagonally, yellow at hoist and red at fly and upper edge.

P is a blue flag with a white rectangle in centre.

Q is a yellow flag.

R is a red flag with yellow St. George's Cross.

S is a is a white flag with blue rectangle in centre.

T is a tricolour flag, red, white and blue, red at hoist, blue at fly.

U is a red and white quartered flag, red at upper hoist and at lower fly.

V is a white flag with a red St. Andrew's Cross.

W is a white flag with a blue border, red rectangle in centre.
X is a white flag with a blue St. George's Cross.
Y is a flag with five yellow bars intersected with five red bars diagonally placed, yellow at upper hoist, red at lower fly.
Z is a black, yellow, blue and red flag, quartered diagonally, black at mast, blue at fly, red beneath, yellow on high.

Code and Answering Pendant. Red and white stripes vertically placed, red at hoist, red at fly.

NUMERAL PENDANTS

No. 1. White with a red circle near the hoist.
No. 2. Blue with a white circle near the hoist.
No. 3. Red at hoist, white in the centre, blue at fly.
No. 4. Red with white St. George's Cross.
No. 5. Yellow at the hoist, blue at the fly.
No. 6. Black upper half, and white lower half.
No. 7. Yellow upper half, and red lower half.
No. 8. White with a red St. George's Cross.
No. 9. White, red, black and yellow quartered, white and red at upper and lower hoist respectively, black and yellow at upper and lower fly respectively.
No. 0. Yellow at hoist and fly, red in centre.

SUBSTITUTES

Substitutes are triangular in shape

First Substitute—Yellow triangle with blue border.
Second Substitute—Blue and white, blue at the hoist, white fly.
Third Substitute—White with black horizontal stripe.

DEFINITIONS

Visual Signalling—Refers to any mode of transmission which is capable of being seen.

Sound Signalling—Refers to the sending of Morse signals by means of the whistle, siren, fog-horn, bell or other sound apparatus.

Message—Denotes any form of communication.

Coded Message—The text consists of figures or letters instead of ordinary words.

Transmitting Ship, or Station—Is the ship (or station) from which the message is actually being made.

Receiving Ship, or Station—Is the ship (or station) which is receiving message.

Procedure—Denotes the rules drawn up for the conduct of signalling.

Procedure Signal—A signal designed to facilitate the conduct of signalling.

Group—Denotes one or more contiguous letters and/or numerals which in themselves compose a separate signal.

Numeral Group—Consists of one or more numerals.

Hoist—Consists of one or more groups displayed from a single halyard.

At the Dip—This is a position of a signal when hoisted about half the extent of the halyards.

Close Up—A signal is close up when hoisted the full extent of the halyards.

Tackline—A length of halyard about 2 metres long. It is used to separate two groups of flags.

CHAPTER II

TYPES OF SIGNALS—SIGNAL LETTERS

A SIGNAL may consist of one, two or three letters (sometimes followed by numbers).

Single-letter Signals

To each letter of the alphabet, is allotted a meaning of special significance. For example **A** signifies 'I have a diver down; keep well clear at slow speed': **B** indicates 'I am taking in or discharging, or carrying dangerous goods': **G** indicates 'I require a Pilot'. A complete list of Single-letter signals is given on page 59. It will be observed that those marked with an asterisk, when made by Sound, may only be made in compliance with the requirements of the International Regulations for Preventing Collisions at Sea, RULES 34 and 35.

Icebreaker signals are also Single-letter signals and can only be used between Icebreaker and assisted vessel, page 62.

Two-letter Signals

Two-letter signals of special importance and which should be committed to memory are:—

- **CB** I require immediate assistance.
- **NC** I am in distress and require immediate assistance.
- **GW** Man overboard. Please take action to pick him up (position to be indicated if necessary).
- **GU** It is not safe to fire a rocket.
- **IT** I am on fire.
- **JL** You are running the risk of going aground.
- **NF** You are running into danger.
- **UY** I am carrying out exercises please keep clear of me.
- **SO** You should stop your vessel instantly.
- **LO** I am not in my correct position. (Signalled by lightvessel.)
- **YG** You appear not to be complying with the traffic regulation scheme.

Three-letter Signals

Three-letter groups cover signals relating to the Medical Section of the *International Code*, and begin with the letter **M**. Extracts from the Medical Section are given on page 108.

Four-letter Signals

Four-letter signals relate to signal letters of ships only. British ships' signal letters begin with **G** or **M**.

Five-letter Signals

The Signal letters for aircraft consist of a group of five letters.

Distress Signals

Distress signals, the importance of knowing which cannot be too strongly emphasised, are given on page 124. They consist of signals made by gun, or other explosive signal, by sound apparatus, radio-telegraphy, rockets or shells, and flags.

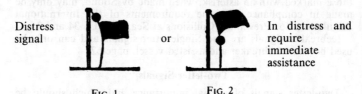

Distress signal   or   In distress and require immediate assistance

Fig. 1   Fig. 2

Signal Letters

Ships: The signal letters of a ship consist of four letters which have been allocated to her for identification purposes. A ship's signal letters are the same as her radio call sign. The nationality of a ship can be determined from the first, or the first two letters.

In the case of ships not fitted with radio, signal letters are allotted to them from the appropriate radio call sign series, as determined by the International Radio-telegraph Regulations. The signals letters of British ships can be found in the publication, *Signal Letters of British Ships*. Four-letter groups beginning with G or M.

Aircraft: The signal letters for aircraft consist of five letters and are the same as their radio call signs. The nationality of the aircraft is determined by the first letter or first two letters of the group. The 'nationality marks' are made into signal letters or registration marks by adding three or four letters to complete the group to five letters.

In civil aircraft the five-letter group is painted on the lower surface of the main plane and also on each side of the fuselage. A hyphen separates the 'nationality mark' from the remaining letters.

A table showing the International allocation of initial letters of signal letters, call signs and aircraft markings is given on page 125.

Signal letters are used to:—
(1) Speak to, or call, a ship, aircraft or station.
(2) Speak of, or indicate, a ship, aircraft or station.

With regard to (1) when speaking to, or calling, a ship the signal letters of the ship being called precede the signal. With respect to (2) when speaking of, or referring to, a ship the signal letters of the ship indicated follow the signal.

For instance, suppose the signal letters of the SS. *Baron Forbes* are GBMR and those of the SS. *City of Exeter* GQZW, and the Code Group CG1 signifies. 'I will stand by to assist you' (or vessel indicated). The signal GBMR—CG1 would mean to *Baron Forbes*, 'I will stand by to assist you'. The signal GBMR—CG1—GJRN would mean to *Baron Forbes*, 'I will stand by *British Captain*'.

With regard to the second signal, GJRN, the signal letters of the *British Captain* is a complementary group and must be signalled after the signal to which it refers.

When a ship's name occurs in the text of a coded message, it is to be expressed by her signal letters.

CHAPTER III

THE USE OF NUMERAL PENDANTS—THE USE OF SUBSTITUTES

THE numeral pendants are, as their designation implies, to be used exclusively for the signalling of numbers. Being pendant-shaped, they are easily distinguished from the alphabetical flags and therefore do not require any additional signal to indicate they represent numbers. The numeral pendants are ten in number, and when used in conjunction with substitutes are available for signalling any number.

When the number being signalled contains a decimal, the answering pendant is to be inserted where it is desired to express the decimal point.

It will be necessary to use the numeral pendants when signalling positions, i.e. latitude and longitude, times, courses and bearings, and distances. The procedure to be adopted in each of those cases will now be explained.

Latitude and Longitude

Latitude—The hoist consists of four numeral pendants preceded and joined by the letter L. The first two figures indicate degrees and the last two minutes, thus L1240 would indicate 12° 40′.

Longitude—Is expressed by four numeral pendants preceded and joined by the letter G, thus G3025 would indicate 30° 25′. Latitude and longitude will require two hoists, the first hoist representing latitude and the second longitude.

Example—L5624, G3025 would indicate Latitude 56° 24′ Longitude 30° 25′.

In visual signalling, it will seldom be necessary to signify North or South Latitude, East or West Longitude.

Occasions may arise in the vicinity of the Equator, the 0° or 180° meridians, when it will be necessary to indicate the hemisphere. In such cases, to avoid confusion, the flags N or S representing North or South, E or W indicating East or West, are, as the circumstances warrant, to be added to the latitude and longitude groups respectively.

Example—L0010N, G0045E would read latitude 00° 10′ North, Longitude 00° 45′ East.

Example—A vessel, by visual signalling, wishing to signal her longitude of 135° 20′ to another vessel would hoist G3520.

It is to be understood, however, that it is permissible to increase the hoists to a five-figure group in circumstances where the recommended form would give rise to confusion.

A numeral group preceded by the letters L or G as shown above is designated a Position Signal.

Times—Times, when signalled, are to be expressed in the 24 hours notation under which system midnight is 00 hours, and the following hours are reckoned continuously up to 23 hours.

Examples—
1.30 a.m. is to be expressed as 0130
10.15 a.m. " " 1015
2.30 p.m. " " 1430
5.45 p.m. " " 1745
11.20 p.m. " " 2320
Midnight " " 0000

Thus, hours and minutes can be signalled by a group of four numerals, the first two numerals indicating hours and the second two minutes. This group is to be preceded and joined by the letters T or Z, forming a single group. Thus, T1430 indicates 1430 hours or 2.30 p.m. Local Time. Z1430 indicates 1430 hours or 2.30 p.m. Greenwich Mean Time.

A numeral group preceded by the letters T or Z, as shown by Plate 3, is designated a Time Signal.

It is sometimes desirable to attach a 'Time of Origin' at the end of a message. Such time should be given to the nearest minute and expressed by four figures. It indicates when the message originated and forms a convenient reference number.

Courses and Bearings—All courses and bearings when being signalled are to be expressed in the three figure notation under which system North is 000° and the remaining degrees are reckoned clockwise continuously up to 359°. Thus, N. 8° E. is 008°, N.8° W is 352°, S.40° E is 140°, S. 80° W. is 260°. Such courses and bearings are to be considered *true* unless, of course, expressly stated to be *magnetic*. For instance, S. 50° W. magnetic would be signalled 230° magnetic.

Courses—A course is to be signalled by a group of three numeral pendants. Such group will be preceded by a letter or group from the code.

In the code, the group MI signifies, 'I am altering course to', hence the hoist MI 125 would read, 'I am altering coure to 125° True'.

Courses are always to be true unless expressly stated to be otherwise in the context.

Azimuth Bearings—When signalling bearings the three numeral pendants may be preceded by, but joined with, the letter A to form a single group.

Thus A125 signifies bearing 125° True.

In the Code the group LT signifies 'Your bearing from me (or from) (name or identity signal) is ... (at time indicated)'.

The hoists LT A125 would read, 'Your bearing from me is 125° True'. Words in brackets are optional and may be omitted when necessary.

Bearings made by a ship pointing out an object or referring to a position are always to be reckoned from the ship making the signal.

Relative bearings is the direction with reference to the fore and aft line of the vessel from which the bearing is taken.

Distances—The distance is to be expressed in miles. If any other unit is used it must be expressed in the context.

Figures preceded by the letter R indicates distances in nautical miles.

When signalling by Code a message which contains numbers, the numeral group must be sent separate from the Code group. This rule, however, does not apply to position signals, time signals, or bearing signals, which are respectively preceded by, and joined with the letters G, L, T, Z and A.

Example—In the Code OV signifies 'Mine—s is/are believed to be bearing ... from me distance ... miles' thus OV-A135-R15 would mean 'Mine—s is/are believed to be bearing from me 135° True distance 15 miles'. The letter R may be omitted if there is no possibility of confusion.

THE USE OF SUBSTITUTES

Substitutes may be employed when the composition of a Code group, or Numeral group, contains a repetition of a letter or figure respectively. For instance, in the code the group DD signifies, 'Boats are not allowed to come alongside'. Without the aid of substitutes, it would be necessary to carry two sets of Code flags before this signal could be made. Again, to signal 1000, three sets of numeral pendants would be required if recourse were not made to substitutes.

The three substitutes will render it possible to signal any combination of four letters or four figures.

In practice most foreign-going ships carry two or more sets of flags.

BROWN'S SIGNALLING

Substitutes are triangular in shape to avoid confusion with alphabetical flags or the numeral pendants. They are to be referred to as first substitute, second substitute and third substitute.

In the International Code of Signals there are two classes of flags, alphabetical flags and numeral pendants. *A substitute is used to repeat a flag of the same class that immediately precedes it.* In other words, if a substitute immediately follows an alphabetical flag, or flags, it repeats one of the flags. If it immediately follows a numeral pendant, or numeral pendants, it repeats one of the pendants.

The following rules are to be adhered to.

1. The **first substitute** always repeats the uppermost signal flag of the class of flags which immediately precedes the substitute.

2. The **second substitute** always repeats the second signal flag counting from the top of that class of flags which immediately precedes the substitute.

3. The **third substitute** always repeats the third signal flag counting from the top of that class of flags which immediately precedes the substitute.

4. No substitute can ever be used more than once in the same group.

5. The answering pendant when used as a decimal point is to be disregarded in determining which substitute to use.

The following examples will illustrate the use of substitutes:—

Example—In the Code book the group MMJ signifies 'Patient has been isolated'. (A)

MAA signifies, 'I require urgent Medical advice'. (B)

T0330 signifies, 'At 0330 Local Time'. (C)

They would be signalled as follows:

(A)	(B)	(C)
Alphabetical Flag M	Alphabetical Flag M	Alphabetical Flag T
First Substitute	Alphabetical Flag A	Numeral Pendant 0
Alphabetical Flag J	Second Substitute	Numeral Pendant 3
		Second Substitute
		First Substitute

In (A) the uppermost flag is to be repeated, hence the first substitute is used, in (C) the first numeral pendant is to be repeated, hence the first substitute is used.

Example—The Signal Letters of the British ship *Estala* are GGGE It would be signalled as follows:—

G
First Substitute
Second Substitute
E

As mentioned before, when signalling numerals involving the use of substitutes, the same rules as apply to alphabetical flags are to be observed.

Example—It is required to signal the number 1001. It is to be signalled as follows:—

    Numeral pendant   1 = 1
    Numeral pendant   0 = 0
    Second substitute   = 0
    First substitute   = 1

The second substitute repeats the second flag counting from the top, in this case numeral pendant 0. The first substitute repeats the first flag of the group, in this case numeral pendant 1.

Example—It is required to signal the following position, Latitude 20° 30′, Longitude 40° 00′. The hoists would be as follows:—

1st Hoist		2nd Hoist	
Alphabetical flag	L	Alphabetical flag	G
Numeral pendant	2	Numeral pendant	4
Numeral pendant	0	Numeral pendant	0
Numeral pendant	3	Second substitute	
Second substitute		Third substitute	

It is to be observed that in each of the above hoists two classes of flags are employed. Bearing in mind, however, that a substitute *repeats a signal flag of the same class as that immediately preceding it,* the substitutes in the above groups refer to the numerals contained therein and not to the alphabetical flag. In the first hoist, the second substitute repeats the second numeral of the hoist, i.e. 0. In the second hoist, the second and third substitutes repeat the second and third numerals respectively in the hoist, i.e. 0 and 0.

Example—To signal 15·5.

    Numeral pendant   1 = 1
    Numeral pendant   5 = 5
    Answering pendant   = · (decimal)
    Second substitute   = 5

Example—To signal 3·55.

    Numeral pendant   3 = 3
    Answering pendant   = · (decimal)
    Numeral pendant   5 = 5
    Second substitute   = 5

Note that the answering pendant, used to express the decimal point, is disregarded when determining which substitute to use.

CHAPTER IV

ALPHABETICAL SIGNALS

The following course is to be adopted whenever names of vessels or places occur in the text of a message being signalled by flags:—

The Signal YZ signifying "The words which follow are in Plain Language' can be used if necessary, or in the case of the signal RV 'You should proceed to', YZ may be omitted.

Example—RV Glasgow. 'You should proceed to Glasgow.'

CHAPTER V

ARRANGEMENT OF THE CODE BOOK
EXAMPLES OF CODING AND DECODING

The arrangement of the *International Code Book* is similar to that on pages 63 to 123 which contain extracts from the Code book.
 1. Single-letter Signals.
 2. Single-letter Signals used in conjunction with numerals.
 3. Single-letter signals to be used when operating with Icebreakers.
 4. Two-letter signals supplemented if necessary by a numeral.
 5. A table of complements which can be used with one- or two-letter signals.
 6. Three-letter signals.
 (A) Single-letter signals are allocated to significations which are urgent, important or of very common use.
 (B) Two-letter signals are from the General Section.
 (C) Three-letter signals beginning with the letter M are from the Medical Section.

The Code follows the basic principle that each signal should have a complete meaning, in certain cases complements are used when necessary to supplement the available groups. Complements express variations in the meaning of the basic signal.

Examples:—

 CN   You should give all possible assistance.
 CNI   You should give immediate assistance to pick up survivors.

EXAMPLES OF CODING

Questions concerning the same basic subject or basic signal.

1. *Examples*:—
 (a) JW   I have sprung a leak.
     JZI   Can you stop the leak?
 (b) Answer to a question or request made by the basic signal.

BROWN'S SIGNALLING

Example:—

IB   What damage have you received?
IB2 I have minor damage.

(1) Supplementary, specific or detailed information.

Examples:—

CB5 I require immediate assistance. I am drifting.
KA   I urgently require a collision mat.

2. Complements appearing in the text more than once have been grouped in three tables. These tables should only be used as and when specified in the text of the signals.

3. Text in brackets indicates:—
   (a) an alternative, e.g. '... (or vessel indicated)...'
   (b) information which may be transmitted if it is required or if it is available, e.g. '...(Position to be indicated if necessary)'
   (c) an explanation of the text.
   (d) The material is classified according to subject and meaning. Extensive cross reference of the signals in the righthand columns is used to facilitate coding.

Example:—

CC5   = I am (or Vessel indicated is) in distress in Latitude... Longitude... (or bearing from place indicated, distance...) and require(s) immediate medical assistance (Complements Table II, if required).
GRBC   = SS. *City of Agra.*
L5745N = Latitude 57° 45′ North.
G0345W = Longitude 03° 45′ West.

Message:—

SS. *City of Agra* in distress in position Latitude 57° 45′ North, Longitude 03° 45′ West, and requires Medical Assistance.

Example:—

JF   = I am (or vessel indicated is) aground in Latitude... Longitude... (also the following complements if necessary).
20 =   2 = Forward
       0 = On Rocky bottom.
L4530N = Latitude 45° 30′ North.
G0352W = Longitude 03° 52′ West.

Message:—

I am aground in position Latitude 45° 30′ North, Longitude 03° 52′ West, forward on rocky bottom.

NOTE:—Whenever the text allows, depths, etc., are to be signalled in feet or metres, the figures followed by 'F' to indicate feet or by 'M' to indicate metres.

KEY TO PLATE 3

No. 1. Position signal: Consists of numeral group preceded by the letter L. It reads Latitude 42° 47′.

No. 2. Bearing signal: Consists of numeral group preceded by the letter A. It reads bearing 355°.

No. 3. Time signal: Consists of numeral group preceded by the Letter T. It reads 1900 Local Time.

No. 4. Ship's signal letters tackline K 9 (Kilo Nine): Signifies that ship displaying hoist desires to communicate by VHF—Channel 16 with vessel whose signal letters are GPNT.

No. 5. Ship's signal letters GRBC; Upper flag being G indicates a British ship.

No. 6. Numeral signals 70·5; Consists entirely of numeral pendants. Answering pendant indicates decimal point.

No. 7. Position Signal: Consists of numeral group preceded by the letter G. It reads Longitude 03° 45′ West.

No. 8. Three-letter signal from the Medical Section, MDG.

No. 9. Two-letter signal NC; Signifies 'I am in distress and require immediate assistance'.

No. 10. Single-letter signal G; Signifies 'I require a Pilot'.

No. 11. Pilot Jack: Hoisted at the fore signifies 'I require a Pilot'.

No. 12. Quarantine signal. For signification *see* page 105.

No. 13. Quarantine signal. For signification *see* page 105.

PLATE No. 4.

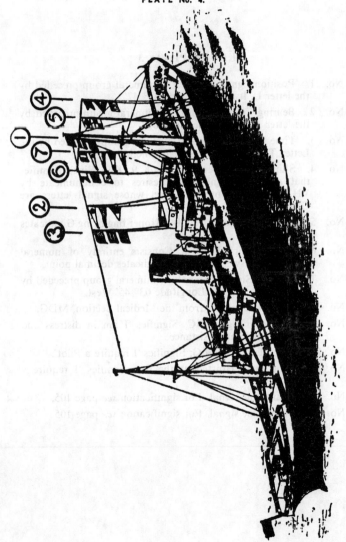

INTERNATIONAL CODE OF SIGNALS
PRINCIPAL TYPES OF HOISTS
PLATE 3

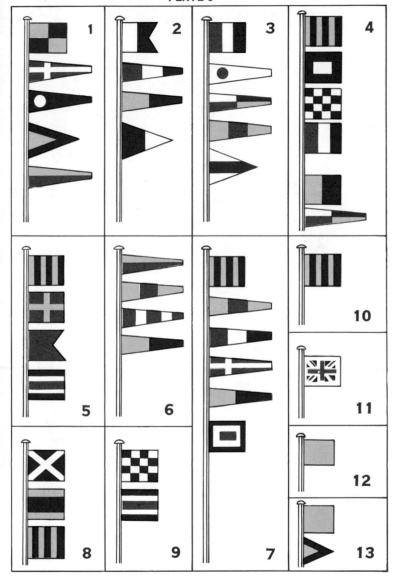

CHAPTER VI

Positions for Hoists—How to Call—How to Answer Signals—How to Complete Signal—When Signals are not Understood—Communication by Local Signal Codes—Communication by Flags between Men-of-War and Merchant Vessels—Questions Relating to Flag Signalling—Answers

When using flags, SIGNALS should always be hoisted in a position where they will be most easily seen by the receiving ship, where they will fly clear and not be obscured, or partially obscured, by smoke or obstruction, in the line of sight.

When it is desired, in order to save time, to display several hoists simultaneously, they may be flown (a) masthead, (b) triatic or fore and aft stay, (c) starboard yardarm, (d) port yardarm, and the sequence in which they are to be read is in the foregoing order.

More than one group may be displayed on the same halyard but they must be separated by a tackline. The upper group is to be read first, second group second, and so on.

Where more than one hoist is shown on the triatic or fore and aft stay, on different halyards, the foremost is to be read first.

In the event of more than one hoist being displayed on the same yardarm, on different halyards, the outboard is to be read first.

Plate 4 shows a ship flying a signal, the different groups forming the signal being displayed simultaneously. The numbers against the groups denote the order in which they should be read.

Generally speaking, only one hoist will be displayed at a time. In every case, however, each hoist, or group of hoists, is to be kept flying until the receiving ship indicates, in the manner described later, that they are understood.

A superior signal is one which has been hoisted before another in respect to time or hoist. Conversely, an inferior signal is one which is hoisted after another in respect of time or hoist.

HOW TO CALL

When signal letters are not hoisted superior to the signal, it will be understood the signal is addressed to all ships within visual signalling distance. In other cases the signal letters of the ship, or ships, addressed are to be hoisted superior to the signal.

When, for any reason, it is impossible to determine the signal letters of the ship it is desired to address, the ship wishing to make the signal should hoist the group VF, which signifies 'You should hoist your identity signal' and at the same time, hoist her own signal letters, and/or the group CS, 'What is the name or identity of your vessel (or station)' could also be used.

If this fails then the group YQ, 'I wish to communicate by...(Complement Table 1) with vessel bearing...from me' must be hoisted.

HOW TO ANSWER SIGNALS

All vessels to which signals are addressed, or are indicated in the signals, are to hoist the answering pendant at the dip as soon as they observe the hoist. As soon as the hoist is understood the answering pendant is to be hoisted to the close-up; it is to be lowered to the dip as soon as the sending ship hauls down her hoist. It is again to be hoisted to the close-up when the next hoist is understood and lowered to the dip when the hoist is hauled down, and so on till the signal is completed.

It is recommended that the triatic stay be not used for the answering pendant in view of the fact that it is sometimes difficult to determine when the pendant is close up in that position.

HOW TO COMPLETE SIGNAL

The transmitting ship is to hoist the answering pendant singly after the last hoist of the signal has been made to indicate the message is completed. The receiving ship will answer in the usual way, i.e. hoist her answering pendant close up.

WHEN SIGNALS ARE NOT UNDERSTOOD

In the event of the receiving ship being unable to distinguish clearly the signals being made to her she is to keep the answering pendant at the dip and hoist an appropriate signal from the Code to inform the transmitting ship the reason of her inability to read the signals.

Again, when she can distinguish the signal but cannot understand the purport of it, she should hoist the signal ZL which signifies 'Your signal has been received but not understood', or the group ZQ, 'Your signal appears incorrectly coded,' you should check and/or repeat the whole.

BROWN'S SIGNALLING

COMMUNICATION BY LOCAL SIGNAL CODES

If a vessel or shore station wishes to make a signal in a local code, then if neccessary, in order to avoid misunderstanding, the following signal from the International Code of Signals should precede the signal.

YVI signifies, 'The groups which follow are from the Local Code'. particulars concerning these local signals should be looked up in the Sailing Directions.

COMMUNICATION BY FLAGS BETWEEN MEN-OF-WAR AND MERCHANT VESSELS

When a man-of-war desires to communicate with a merchant vessel, she will hoist the Code pendant in a conspicuous position and keep it flying during the whole time the signal is being made.

QUESTIONS RELATING TO FLAG SIGNALLING

1. When is the answering pendant at the 'dip'?
2. What is a hoist?
3. What does the Red Burgee hoisted singly indicate?
4. What is the nature of a two-flag signal?
5. You observe a four-flag signal G uppermost. What type of signal is it?
6. When is a signal 'close-up'?
7. What do you understand by a ship's 'signal letters'?
8. When speaking to a ship, what is the position of her signal letters relative to the signal made?
9. When are the signal letters of a ship hoisted inferior to a signal?
10. How would you ascertain the nationality of a vessel from her signal letters?
11. How would you express 2hrs. 30 mins. p.m. Local Time in a signal?
12. Hoists are flying at the triatic stay, masthead and both yardarms. In what order should they be read?
13. When is a signal said to be superior to another?
14. There are two hoists on the same yardarm but on different halyards. Which is to be read first?
15. A vessel is flying a signal without signal letters superior to it. To whom is the signal addressed?
16. What is the objection to using the triatic stay for the answering pendant?

17. Two hoists are suspended from the triatic stay. Which is to be read first?
18. How would you indicate the end of the message?
19. In what respect do the signal letters of ships differ from those of aircraft?
20. You observe a signal being made to you. What would you do?
21. How would you call a vessel?
22. How would you call a vessel whose signal letters you cannot determine?
23. Give the meaning of the hoist Z1530?
24. What is the signification of hoist—T15 1st sub. 2nd sub?
25. In the Code some words have a qualification, in brackets, alongside. What is the object of inserting the qualification?
26. In what order should latitude and longitude be signalled?
27. What is a Tackline, and for what purpose is it used?
28. How would you recognise latitude and longitude hoists?
29. How is the decimal point in a number indicated?
30. What does the hoist A140 indicate?
31. What are quarantine signals?
32. In what manner are Icebreaker signals made?
33. What is the signification of flags YP9 hoisted inferior to a group of signal letters? (*See* Supplements Table.)
34. How does a man-of-war indicate she wishes to communicate by semaphore with a merchant vessel?
35. What is the use of substitutes?
36. Which Flag does the second substitute repeat?
37. Can the same substitutes be used twice in one group?

Code the following—

38. I cannot refloat without jettisoning everything movable aft. Will you escort me to Latitude 5745N, Longitude 0344W after refloating.
39. I have broken adrift. I require 4 tugs.
40. My position at 1430 G.M.T. was Latitude 33° 30′ N., Longitude 44° 40′ W.
41. There is a raft bearing 130° True 2½ miles from me.
42. Have you sighted or heard of a vessel in distress bearing 225° True, 10 miles from Wolf Rock.
43. Vessel coming to your rescue is steering course 225°, Speed 21 knots.
44. My course is 180°.
45. The depth at High water here is 24¼ feet.
46. Your bearing from me is 225° True at 2.30 p.m. Local Time.

47. What is the depth of water over the Bar at 1200 G.M.T.?
48. You should navigate with caution. Small fishing boats are within 12 miles from me.
49. Fishing gear has fouled my propeller.
50. I wish to communicate with you in English.
51. I can communicate with you in French.

Decode the following—

52. GPVR *(Royal Lion)*—IT2.
53. AF1.
54. FL180.
55. BD 1340.
56. BW–031st Sub–Z132nd Sub 0.
57. CD 5.
58. KD 2.
59. PL 3.
60. CF–A 180.
61. GWCP (*Wrestler*) CH–L5742ndSubN–G03 2nd Sub 1st Sub W.
62. OV–A2 1st Sub 5–R2 1st Sub.
63. RV–London.
64. K8.
65. XX 5.

ANSWERS

1. When it is hoisted about half the extent of the halyards.
2. A hoist consists of one or more groups displayed from the same halyard.
3. I am taking in, discharging, or carrying explosives.
4. Contains general section of the code.
5. It is the Signal letters of a British ship.
6. When hoisted the full extent of the halyard.
7. Letters, four in number, which have been allotted to her for identification purposes.
8. Her signal letters are hoisted superior to the signal.
9. When speaking of, or referring to, a ship.
10. Her nationality may be determined from the first letter, or first two letters, of her signal letters.
11. T1430.
12. (i) Masthead, (ii) triatic stay, (iii) starboard yardarm, (iv) port yardarm.
13. When hoisted before another in respect to time or position.
14. The outboard hoist.

15. To all vessels within visual signalling distance.
16. It is difficult for the other ship to discern whether or not it is close-up.
17. The foremost.
18. By hoisting the answering pendant singly after the last hoist.
19. A ship's signal letters are four in number, and the aircraft's signal letters consists of five letters.
20. On observing the hoist, answering pendant to be hoisted at the dip. When hoist is understood, hoist answering pendant close-up.
21. By hoisting her signal letters superior to the signal.
22. Request the vessel to hoist her signal letters by hoisting the appropriate group from the code. If this fails, then hoist the group indicating 'I wish to communicate with vessel on bearing... indicated from me'.
23. It is a Time Signal—15 hours 30 minutes G.M.T.
24. It is a Time Signal—15 hours 15 minutes local Time.
25. The qualification indicates the sense in which the word is to be used.
26. The first group latitude, the second group longitude.
27. A tackline is a 2 metre length of line. It is used to separate each group of flags which, if not separated, would convey a different meaning to that intended.
28. Such hoists consist of four numeral pendants preceded by and joined with the letters L or G.
29. Flag Signalling. By inserting the answering pendant where it is desired to express the decimal point.
30. Bearing 140° True.
31. Signals shown on vessels requiring or required to show their state of health.
32. By any visual means, sound or radio-telephony.
33. It indicates that the vessel flying the signal wishes to communicate by VHF Channel 16 with the vessel whose signal letters are hoisted.
34. She will hoist the Code pendant in a conspicuous position and the signal letters of the merchant vessel.
35. To enable the same flag to be repeated one or more times in the same group while still only carrying one set of flags.
36. The second substitute always repeats the second signal flag counting from the top of that class of flags which immediately precedes the substitute.
37. No substitute can ever be used more than once in the same group.

38. JQ4–JV–L5745N–G034 3rd Sub W.
39. RC1–KF4.
40. ET–Z1430–L3 1st Sub 2nd Sub 0 N
    –G4 1st Sub 2nd Sub 0 W
41. DP—A130–R2 Code Pt5.
42. EA–A2 1st Sub 5–Wolf Rock–R10.
43. GR–C2 1st Sub 5–S21.
44. MD 180.
45. QA–24 Code Pt 1st Sub 5–F.
46. LT–A2 1st Sub 5–T1430.
47. LRI–Z1203rd Sub.
48. TH–12.
49. TP.
50. ZA1.
51. ZB2.
52. Vessel *Royal Lion* is on fire.
53. Do you intend to abandon your vessel?
54. You should steer course 180° to reach position of accident.
55. I have established communication with aircraft in distress on 1340 kHz.
56. The magnetic course for you to steer towards me is 030° at 1330 GMT.
57. I require medical assistance.
58. There are two compartments flooded.
59. You can pass the buoy on either side.
60. Signals from Vessel/Aircraft requesting assistance are coming from bearing 180° from me.
61. Vessel *Wrestler* is reported as requiring assistance in Latitude 57° 47′ North, Longitude 03° 30′ West.
62. Mine(s) is believed to be bearing 225° from me distance 22 miles.
63. You should proceed to London.
64. I wish to communicate by Radiotelephony on 2182 kHz.
65. True direction of the wind is Southwest.

CHAPTER VII

SEMAPHORE SIGNALLING

SEMAPHORE signalling is a very convenient mode of communication. The signals may be made by means of the mechanical semaphore, or by the signalman with or without small flags and forming the different letters by moving his arms. In either case great care must be exercised to ensure the letters and signs are formed accurately, otherwise delay will arise.

Although not now examined for certificates of competency, this chapter is retained for use by Scouts, Guides or others who may want to learn semaphore signalling. The Semaphore alphabet is given on page 38.

A station which desires to communicate with another station by semaphore may indicate the requirement by transmitting to that station K1 (Kilo One) by any method. If the stations are close to one another the attention sign may be made instead.

All messages transmitted by semaphore are to be made in plain language; should numbers occur in the message they are to be spelt out in words. A decimal point between numerals is spelt out, thus 'DECIMAL'.

Procedure for Sending and Receiving

The sender faces the station addressed and makes the attention sign. The station addressed on observing it makes the answering sign. The sender after a short pause carries on with the message.

The end of a word is indicated by dropping the arms to the break position. When double letters occur in a word after the first of the double letters is made, the arms are dropped to the break position and then without a pause moved to the second letter.

The receiver will acknowledge the correct reception of each word by making the letter C. If this letter is not made, the sender repeats the word.

should an error occur, it is to be indicated by a succession of E's (EEEEE, etc.). The last word correctly transmitted should now be made and the message continued. The ending sign AR is to be made at the end of the message.

BROWN'S SIGNALLING

A	B	C ANSWERING SIGN	D	E
F	G	H	I	J
K	L	M	N	O
P	Q	R	S	T
U	V	W	X	Y
	Z	ATTENTION	BREAK	

Example—Station 'A' wishes to ask, by semaphore, if Station 'B' can supply two coils of three-inch manila rope.

The procedure is as follows:—

Component	Station 'A' makes	Station 'B' makes
Call	Attention Sign	Answering Sign C
Identity	Station 'B' de Station 'A'	C
Text	Can you supply me with two coils of three inch manila rope	C C C C C C C C C C C C
Ending	AR	R

HOW TO LEARN SEMAPHORE

(*See* Diagram on page 35)

First Circle: Make the letters A, B, C, D, E, F, G. The letter D can be made with either hand.
Second Circle: Keeping the RIGHT HAND at 'A' position and moving the LEFT HAND only to H, I, K, L, M, N. (Omit the letter J.)
Third Circle: Keeping the RIGHT HAND at 'B' position and moving the LEFT HAND only to O, P, Q, R, S.
Fourth Circle: Keeping the RIGHT HAND at 'C' position, move the LEFT HAND only to T, U, Y.
Fifth Circle: Keeping the RIGHT HAND at 'D' position, move the LEFT HAND only to J, V.
Sixth Circle: Keeping the LEFT HAND at 'E' position, move the RIGHT HAND only to W, X.
Seventh Circle: Keeping the LEFT HAND at 'F' position move your RIGHT HAND only to Z.

NOTE—'First Circle'—The letters E, F and G must be made with the LEFT HAND.

CHAPTER VIII

MORSE SIGNALLING—EXPLANATION OF USE OF PROCEDURE SIGNALS AND SIGNS—FORM OF MESSAGE—EXAMPLES OF TRANSMISSIONS OF MESSAGES—RATE OF SIGNALLING AND PERCENTAGE TABLES.

Morse signalling is carried out by flashing a light, or by sound. The symbols which represent letters are expressed by two elements called a dot (or short) and a dash (or long), signalled singly or in combination. The following ratios are to be observed between dot, dash, letter, word or group.

A dot is taken as the unit and a dash is equivalent to three units.

Interval between each flash or sound .... 1 unit
Interval between each letter .... 3 units
Interval between each word or group .... 7 units

The above spacing should be adhered to whatever the rate of sending.

The standard rate of signalling by flashing in Morse is to be regarded as six words per minute. The International Morse Code alphabet, numerals, procedure signals and signs are given on pages 39 and 41, and should be committed to memory.

The following Tables give a list of the Morse symbols used for visual and sound signalling.

INTERNATIONAL MORSE CODE

Alphabet

Meaning	Symbol	Meaning	Symbol	Meaning	Symbol
A	·−	J	·−−−	S	···
B	−···	K	−·−	T	−
C	−·−·	L	·−··	U	··−
D	−··	M	−−	V	···−
E	·	N	−·	W	·−−
F	··−·	O	−−−	X	−··−
G	−−·	P	·−−·	Y	−·−−
H	····	Q	−−·−	Z	−−··
I	··	R	·−·		

Numerals

Meaning	Symbol
1	·−−−−
2	··−−−
3	···−−
4	····−
5	·····
6	−····
7	−−···
8	−−−··
9	−−−−·
0	−−−−−

Punctuation

Meaning	Sign	Symbol
Full stop and Decimal Point	AAA	·−·−·−

HOW TO LEARN THE MORSE CODE
(*See* page 39)

Learn in sequence of columns as indicated below:

1. *E, I, S, H, T, M, O.*
2. *A, U, V.*
3. *N, D, B.*
4. *A, W, J.*
5. *C, K, P, G.*
6. *R, L, Q, Z.*
7. *F, X, Y.*
8. Figures 1 to 0.

Learn one column at a time before proceeding to the next column, then practice regularly.

Procedure Signals and Signs

Meaning	Sign	Symbol
Call for unknown ship and general call	*AA AA*, etc.	▄▄ ▄ ▄▄ ▄▄ ▄ ▄▄ etc.
Answering sign	*TTTTT*, etc.	▄▄ ▄▄ ▄▄ ▄▄ ▄▄
Erase sign	*EEEEEEEE*	▄ ▄ ▄ ▄ ▄ ▄ ▄ ▄ etc.
Repeat sign	*RPT*	▄ ▄▄ ▄ ▄ ▄▄ ▄▄ ▄ ▄▄ ▄
All after	*AA*	▄ ▄▄ ▄ ▄▄
All before	*AB*	▄ ▄▄ ▄▄ ▄ ▄ ▄
Word or group between	*BN*	▄▄ ▄ ▄ ▄ ▄▄ ▄
Word or group after	*WA*	▄ ▄▄ ▄▄ ▄ ▄▄
Word or group before	*WB*	▄ ▄▄ ▄▄ ▄▄ ▄ ▄ ▄
Ending sign	*AR*	▄ ▄▄ ▄ ▄▄ ▄
From	*DE*	▄▄ ▄ ▄ ▄
Yes, or the signification of the previous group should be read in the affirmative	*C*	▄▄ ▄ ▄▄ ▄
Message received	*R*	▄ ▄▄ ▄
Word (plain language) received	*T*	▄▄
Code groups follow	*YU*	▄▄ ▄ ▄▄ ▄▄ ▄ ▄ ▄▄

SIGNALS FOR USE WHERE APPROPRIATE IN ALL FORMS OF TRANSMISSION

AA   Used on conjunction with RPT sign (All after).

AB   Used in conjunction with RPT sign (All before).

AR	End of message sign.
AS	Waiting signal.
BN	Used in conjunction with the RPT sign (All between).
C	Affirmative 'The signification of the previous group should be read in the affirmative'.
CS	'What is the name or identity signal of your vessel (or station).'
De	'From...' (used to proceed the name or identity signal of the calling station)
K	'I wish to communicate with you' or 'Invitation to transmit.'
NO	Negative 'The significaton of the previous group should be read in the negative'. In voice pronounced NO.
OK	Acknowledging a correct repetition 'It is correct'.
RQ	Interrogative, or 'the significance of the previous group should be read as a question'.
R	Received or 'I have received your last signal'.
RPT	Repeat sign 'I repeat' or 'Repeat what you have sent', or 'repeat what you have received'.
WA	Used in conjunction with RPT sign (Word after).
WB	Used in conjunction with RPT sign (Word before).

NOTES—The signals C, NO, and RQ

    (a) are NOT to be used in conjunction with single-letter signals.

    (b) when used by voice they should be pronounced in accordance with the Letter-spelling Table.

LETTER-SPELLING TABLE

NOTE—The syllables to be emphasised are italicised.

Letter	Codeword	Pronounced
A	Alfa	*AL*-FAH
B	Bravo	*BRAH*-VOH
C	Charlie	*CHAR*-LEE
D	Delta	*DELL*-TAH
E	Echo	*ECK*-OH
F	Foxtrot	*FOKS*-TROT
G	Golf	GOLF
H	Hotel	HOH-*TELL*
I	India	*IN*-DEE-AH
J	Juliet	*JEW*-LEE-ETT
K	Kilo	*KEY*-LOH
L	Lima	*LEE*-MAH
M	Mike	MIKE
N	November	NO-*VEM*-BER
O	Oscar	*OSS*-CAH
P	Papa	PAH-*PAH*
Q	Quebec	KEH-*BECK*
R	Romeo	*ROW*-ME-*OH*
S	Sierra	SEE-*AIR*-RAH
T	Tango	*TANG*-GO
U	Uniform	*YOU*-NEE-FORM (or *OO*-NEE-FORM)
V	Victor	*VIK*-TAH
W	Whisky	*WISS*-KEY
X	X-Ray	*ECKS*-RAY
Y	Yankee	*YANG*-KEY
Z	Zulu	*ZOO*-LOO

FIGURE-SPELLING TABLE

Fig. or mark to be transmitted	Codeword	Pronounced
0	Nadazero	NAH-DAH-ZAY-ROH
1	Unaone	OO-NAH-WUN
2	Bissotwo	BEES-SOH-TOO
3	Terrathree	TAY-RAH-THREE
4	Kartefour	KAR-TAY-FOWER
5	Pantafive	PAN-TAH-FIVE
6	Soxisix	SOK-SEE-SIX
7	Setteseven	SAY-TAY-SEVEN
8	Oktoeeight	OK-TOH-AIT
9	Novenine	NO-VAY-NINER
Decimal Pt.	Decimal	DAY-SEE-MAL
Full Stop	Stop	STOP

NOTE—Each syllable is equally emphasised. The second component of each Code word is the Code word used in the Aeronautical Mobile Service.

SIGNALS FOR VOICE TRANSMISSION (RADIO-TELEPHONY OR LOUD HAILER)

Signal	Pronounced	Meaning
Interco	IN-TER-CO	International Code groups follow.
Stop	STOP	Full Stop
Decimal	DAY-SEE-MAL	Decimal Point
Correction	KOR-REK-SHUN	Cancel my last word or group. The correct word or group follows.

SIGNALS FOR FLAGS RADIOTELEPHONY AND RADIOTELEGRAPHY TRANSMISSIONS

CQ   Call for unknown station/s or General call to all stations.

NOTE—When this signal is used in voice transmissions it should be pronounced with the letter-spelling table.

EXPLANATION OF THE USE OF PROCEDURE SIGNALS AND SIGNS

The use of procedure signals and signs is to govern the form or mode of exchanging messages between ships and, at the same time, to provide means by which signals can be efficiently, yet rapidly, transmitted with accuracy.

AA   AA   AA, etc.

The Call Sign—This is the normal method of calling up at sea. It is to be used to call an unknown ship and as a General call, and is to be continued until the ship addressed answers.

TTTTTTT, etc

This sign, a succession of T's, is the Answering sign to the call, and is to be made until the transmitting ship ceases to call.

AS

Wait Sign—This sign when made to a vessel or station means wait.

K

Transmit Sign—Signifies 'You may now transmit' or 'Instructions to transmit'.

EEEEEEEE

The Erase Sign—A succession of E's indicates that the last word or group was signalled incorrectly. It is to be answered with the erase sign. As soon as it has been answered, the transmitting ship will repeat the last word or group which was correctly signalled then carry on with the remainder of the message.

In the event of the mistake not being discovered until after the complete message has been signalled, a new message must be made.

If, while in the process of transmission, it is desired to cancel the whole message, the erase sign is to be made followed by the ending sign AR, viz. EEEEEEEEEE AR.

RPT

The Repeat Sign is used to obtain a repetition of the whole or part of a message.

When made singly is signifies 'Repeat the last message'. In this case, the repetition is signalled, making the message in exactly the same form as originally transmitted.

On many occasions it will not be neccessary to request a repetition of the whole message, in which case the repeat sign is to be used in conjunction with one of the signs AA AB BN WA WB, whichever is appropriate.

RPT AA SIGHTED	signifies	'Repeat all after the word SIGHTED'.
RPT AB MCV	signifies	'Repeat all before the group MCV'.
RPT BN Proceed-boat	signifies	'Repeat all between Proceed and Boat'.
RPT WA LOAD	signifies	'Repeat the word after LOAD'.
RPT WB YS	signifies	'Repeat the group before YS'.

It is to be noted that when a message is not understood or a decoded message not intelligible, the repeat sign is NOT to be made. In such cases the receiving ship must make the appropriate signal from the code book.

NOTE—In Sound signalling, the repeat sign when made singly has a special signification. *See* page 56.

AR

The ENDING SIGN AR is used in all cases to end a message and answered with R.

De

The word 'De' used in the identity signifies 'From—' Thus: De GQMJ indicates 'From a ship whose signal letters are GQMJ'.

C

The letter C signifies 'Affirmative', the signification of the previous group should be read in the affirmative.

R

The letter signifies 'Message received'.

T

This letter is used to indicate the receipt of each word in the text of a plain language message.

YU

This is a Code group signifying 'I am going to communicate with your vessel or station by means of the International Code of Signals'.

In messages transmitted by the Morse Code 'YU' is to be used as the first group to indicate that groups which follow are from the International Code and is not a plain language message. Receipt of each group is acknowledged by the letter T.

FORM OF MESSAGE

A message made by flashing will, in general, consist of the following components, although all of these components are not necessarily signalled in every message:
  1. Call.
  2. Identity.
  3. Text.
  4. Ending.

The Call—the transmitting ship makes the call and continues doing so until answered. The call consists of:—

 (i) The general call (AA AA AA) etc., or
(ii) The signal letters of the ship called.

The answering ship, on observing the call and when ready to read and write down, will answer by making the answering sign TTTTT etc.

The Identity—It will not always be necessary for ships to exchange identity, but should it be required to do so the following is the procedure.

On the call being acknowledged, the transmitting ship will make *de* (from)' followed by her signal letters. This will be repeated back.

The receiving ship will now make her own signal letters, which the transmitting ship will repeat back.

In the event of either ship failing to repeat back immediately, or repeat back correctly, the other ship will make her signal letters again until they are correctly repeated back.

The Text—The text refers to the message being signalled, and consists of words in plain language or Code groups. Each word or group is signalled separately. The receiving ship will:—

(a) Acknowledge the receipt of each word or group with the letter T.
(b) If the receiving ship does not acknowledge a word or group, the transmitting ship should immediately signal again the last word or group.

Ending—At the conclusion of the text the transmitting ship makes the ending sign (AR). The ending is answered by R.

When two ships are signalling for a considerable period and several messages are passed between them, the call and identity need be signalled in the first message only, in order to avoid delay.

Numbers will usually be signalled by the numerals in the Morse Code, but they may be spelt out in a plain language message. Thus 680 may be written and transmitted as 'six eight zero'.

EXAMPLES OF TRANSMISSION OF MESSAGES

With the object of illustrating the use of the procedure signals and signs, it will be advisable to give here a few examples of the transmission of messages between ships.

Example 1. **Plain Language Message**

The SS. *Euripides* (signal letters GMLP) signals to the SS *City of Calcutta* (signal letters GCPK) the message, 'Is the monsoon blowing strong?' The arrows show the direction of transmission.

Component	SS. *Euripides* makes	SS. *City of Calcutta* makes
Call	AA AA AA, etc. →	TTTTT, etc.
	De GMLP →	GMLP
Identity	GCPK ←	De GCPK
	Is →	T
	the →	T
Text	monsoon →	T
	blowing →	T
	strong →	T
Ending	AR →	R

Example 2. **Coded Message**

The SS. *Loch Leven* (signal letters GSXZ) asks the SS. *Mateba* (signal letters OTBA) the question: 'Is vessel SS. *Hilary* (signal letters GQVM) in distress?'

Component	SS. *Loch Leven* makes	SS. *Mateba* makes
Call	AA AA AA, etc. →	TTTTT, etc.
Identity	De GSXZ →	GSXZ
	OTBA ←	De OTBA
Groups from Int. Code	YU →	T
Text	DZ1 →	T
	GQVM →	T
Ending	AR →	R

NOTE—Words or Plain language may also be used to signal names or places, etc.

Receipt of each word or group is acknowledged by the letter T.

Example 3. **Requesting Repetition**

The SS. *Southern Star* (signal letters WGEB) has signalled to the SS. *Northern Light* (signal letters KGEG) the message 'Sighted iceberg in position L4130 G0720'. The SS. *Northern Light* requests a repetition of all after position.

The signalling is carried out as follows:—

Component	SS. *Northern Light* makes	SS. *Southern Star* makes
Call	WGEB, etc. →	TTTTT, etc.
Identity	De KGEG →	KGEG
	RPT AA Position →	T
Ending	AR →	R

The SS *Southern Star* then repeats 'Position' L4130, G0720 as follows:—

Component	SS. *Southern Star* makes		SS. *Northern Light* makes
Call	KGEG, etc.	→	TTTTT, etc.
Identity	De WGEB	→	WGEB
Text	Position	→	T
	L4130	→	T
	G0720	→	T
Ending	AR		R

When requesting the repetition, and on signalling the repetition, the call and identity signs could have been omitted since the vessels were previously signalling.

Example 4. **Coded Message in which two mistakes are made in transmission**

SS. *Elmpark* (signal letters GDMF) transmits to SS. *Daylight* (signal letters WDEJ) a coded message signifying 'I have located wreckage from SS. *Nebula* in distress in position Latitude 57°45′ North, Longitude 03°45′ West'.

Component	SS. *Elmpark* makes		SS. *Daylight* makes
Call	AA AA AA, etc.	→	TTTTT, etc.
	De GDMF	→	GDML
	De GDMF	→	GDMF
	WDEJ	←	De WDEJ
	YU	→	T
	GL	→	T
	GNYA	→	T
Text	L5740N	→	T
	EEEEEEE	→	EEEE
	GNYA	→	T
	L5745N	→	T
	G0345W	→	T
Ending	AR	→	R

NOTE—Observe that *Daylight* repeated back incorrectly the group GDMF, and *Elmpark* again made her signal letters.

In the text *Elmpark* made L5740 instead of L5745. Discovering the error, she made the erase sign which *Daylight* acknowledged. *Elmpark* then made GNYA the last group signalled, and carried on with the remainder of the message.

Example 5. **Requesting Repetition**

The SS. *Earl Haig* (signal letters GQMJ) has signalled to the SS. *City of Agra* (signal letters GRBC) in Plain language the message 'Understand Willapa buoy is adrift'. The SS. *City of Agra* requests a repetition of the word before 'buoy'. The signalling is carried out as follows:—

Component	SS. *City of Agra* makes	SS. *Earl Haig* makes
Call	GQMJ, etc →	TTTTT, etc.
Identity	De GRBC →	GRBC
Text	RPT WB Buoy →	T
Ending	AR →	R

The SS. *Earl Haig* then repeats the word required, and the word quoted, as follows:—

Component	SS. *Earl Haig* makes	SS. *City of Agra* makes
Call	GRBC, etc. →	TTTTT, etc.
Identity	De GQMJ →	GQMJ
Text	Willapa →	T
	Buoy →	T
Ending	AR →	R

NOTE—The Call and identity could of course, have been omitted.

IANS/DTp
EXAMINATIONS IN SIGNALLING

A new signals examinations procedure, authorised by the Department of Transport (DTp) and conducted by colleges of the International Association of Navigation Schools (IANS), commenced on 1st January, 1987.

The examinations apply, as appropriate, to candidates for Class 2,

Class 4, Class 5 and Fishing Certificates of Competency. They will consist of two parts—a reading of morse flashing and an Oral/Practical part.

MORSE FLASHING

A block of sixty (60) characters will be transmitted in 3 minutes for examinations requiring competence at 6 words per minute (wpm) and in 4 minutes for examinations requiring competence at 4 wpm. These timings incorporate an allowance for the greater number of symbols involved in numerals.

The characters will represent a good mixture of letters and numerals. There may, for example, be one numeral in every line although this is not a requirement. The procedural signal AR will be sent after the sixty characters.

The block will be followed by a suitable message of ten words, averaging 5 letters per word as far as possible. This will be sent in 1 minute 40 seconds (for 6 wpm) or in 2 minutes 30 seconds (for 4 wpm).

The scoring for the above will be 1 mark per correct character for the block and 4 marks per complete correct word for the message. The total possible score is therefore 100. For a pass, 90% overall must be scored.

ORAL/PRACTICAL TEST

Questions may cover
a) Morse procedures
b) International Code flag recognition, single letter and selected important two-letter signals.
c) Competent use of the International Code Book.

In addition the candidate may be asked to send morse by key.

Radio D.F. operation and Lifeboat radio operation are not now included in the signals examinations.

Candidates for Class two may also be asked Emergency Telegraphy transmission procedure.

CHAPTER IX

MORSE SIGNALLING BY HAND-FLAGS OR ARMS

1. Raising both hand-flags or arms.

2. Spreading out both hand-flags or arms at shoulder level.

3. Hand-flags or arms brought before the chest.
   Separation of "dots" and/or "dashes".

4. Hand-flags or arms kept at 45° away from the body downwards.
   Separation of letters, groups or words.

5. Circular motion of hand flags or arms over the head.
(a) Erase signals, if made by the transmitting station.
(b) Request for repetition if by receiving station.

Note:—The space of time between dots and dashes and between letters, groups or words should be such as to facilitate correct reception.

1. A STATION which desires to communicate with another station by Morse signalling by Hand-flags, or Arms, may indicate the requirement by transmitting to that station the signal K2 (KILO TWO) by any method.

The Call signall AA AA AA may be made instead.

2. On receipt of the call that station addressed should make the answering signal, or, if unable to communicate by this means, should reply with the signal YS2 by any available method.

3. The **Call Signal** AA AA AA and the signal T should be used respectively by the transmitting station and the addressed stations.

4. Normally both arms should be used by this method of transmission, but in cases where this is difficult or impossible, one arm can be used.

5. All signals end with the Ending Sign AR.

CHAPTER X

RADIOTELEPHONY (R/T)

1. When using the International Code of Signals, in cases of language difficulties, letters are to be spelt out in accordance with the Spelling Table.

2. When stations and ships are called the identity signals (call signs) or name shall be used.

3. **Calling**—The Call consist of:—
The call sign or name of station/vessel called not more than three times at each call.
The group De (Delta Echo).
The call sign or name of the calling station not more than three times at each call.

Difficult names of stations should be spelt. After contact has been established, the call sign or name need not be sent more than once.

4. **Replying**—the reply to calls consists of:—
The call sign or name of the calling station not more than three times.
The group De (Delta Echo).
The call sign of name or station called not more than three times.

Calling all stations in the vicinity the group CQ (Charlie Quebec) should be used but not more than three times at each call.

5. To indicate that Code groups from the International Code Book follows the word 'INTERCO' is to be used. Words of Plain language may also be used in the text when the signal includes names, places, etc.

6. If the station called is unable to accept traffic immediately the signal AS (Alfa Sierra) adding the waiting time in minutes whenever possible.

Example—Alfa Sierra 5, 'Wait 5 Minutes'.

7. The Receipt of a message is indicated by R (Romeo).

8. If the message is to be repeated the signal RPT (Romeo Papa Tango) should be used. Supplemented if necessary by the groups AA...All after AB...All before BN...All between WA...Word after WB...Word before.

9. The End of Message is indicated by AR (Alf Romeo).

CHAPTER XI

SOUND SIGNALLING—QUESTIONS RELATING TO MORSE SIGNALLING—ANSWERS

MORSE signalling can be carried out at sea by sound through the medium of whistle, siren, fog-horn, etc. It is of necessity, a rather slow, tedious process and hence the length of the signal should be reduced as much as possible. The misuse of this form of signalling is liable to cause serious confusion at sea, and for this reason should be used with discretion. Masters are reminded that the One-letter signals of the Code, which are marked * when made by Sound may only be made in compliance with the requirements of the *International Regulations for Preventing Collisions at Sea.* Reference is also made to the Single-letter signals, providing for exclusive use between icebreaker and assisted vessels.

In sound signalling the procedure is as follows—As in flashing, the transmitting ship makes the call sign (AA AA, etc.). The receiving ship answers with the answering sign TTTTT, etc. The transmitting ship now carries on with the message right through to the end, when she makes the ending sign AR. When the ending sign has been made the receiving ship makes R message received).

It is to be observed that the receiving ship does not answer (until the ending sign is made. In the event of her missing a word she is immediately to make the Repeat sign RPT when the transmitting ship will cease signalling, then go back a few groups or words and carry on with the message. Thus in Sound singalling, the repeat sign made singly signifies 'I have missed the last word or group, please go back a few words or groups and continue the message'.

Although the general call sign is used the transmitting and receiving ships do not exchange identities.

Example—A vessel which has just passed a floating mine hears the fog-signal of another vessel and wishes to inform her of the presence of the mine. The transmission is carried out as follows:—

Component	Transmitting Ship makes	Receiving Ship makes
Call	AA AA AA, etc →	TTTTT, etc.
Text	{ Have just passed floating mine	
Ending	AR →	R

QUESTIONS RELATING TO MORSE SIGNALLING

1. How would you call a vessel?
2. How would you acknowledge the receipt of a Plain language word?
3. What is the ending sign and how would you answer it?
4. What is the International Code Group Indicator?
5. What sign will the transmitting ship make on your repeating back correctly a Code group?
6. What is the ratio between a Dot and a Dash?
7. What is the space of time, in units, between two words or groups?
8. The Receiving ship repeats back incorrectly the signal letters of the transmitting ship. What should the latter ship do?
9. In Sound signalling when making single-letter signals, is it necessary to make the call and and answering signs?
10. In Sound signalling the receiving ship misses a word. What should she do?
11. During fog, is it necessary for ships signalling by Sound to exchange identities?
12. What is the signal for a Pilot by Morse signalling?
13. When transmitting by Morse and you make a word or group incorrectly what would you do?
14. If the Ending sign has not been made and you wish to cancel a message what would you do?
15. How would you exchange identities with another ship?

Answers

1. By (i) The General Call (AA AA AA, etc.), or
   (ii) The Signal letters of ship to be called.
2. By making the letter T.
3. The Ending sign is AR. Made as one symbol. It is answered by R.
4. The group YU.
5. The letter OK.
6. As one is to three.
7. Seven units.
8. Make her signal letters until they are repeated back correctly.
9. No.
10. Make the Repeat sign RPT.
11. No.
12. The International Code signal G or P.
13. Make the erase sign (EEEEEEE) and go back and start from the last word or group made correctly.
14. Send the erase sign (EEEEEEE) followed by AR (ending sign).
15. Call AA AA answered by T

De GVST	,,	GVST
GTST	←	De GTST
AR	→	R

CHAPTER XII

CODE GROUPS

Single-letter signals—Icebreaker Tables—General Code—Single-letter signals with complement. Complement Tables. Distress signals—Pilot Signals—Quarantine signals—Table showing International allocation of initial letters of Signal letters—Call signs and Aircraft markings—signal letters.

SINGLE-LETTER SIGNALS

May be made by any method of signalling for those marked *. *See* notes.

A   I have a diver down. Keep well clear at slow speed.

*B   I am taking in, or discharging, or carrying dangerous goods.

*C   Yes (Affirmative or 'The signification of the previous group should be read in the affirmative').

*D   Keep clear of me, I am manoeuvring with difficulty.

*E   I am altering my course to starboard.

F   I am disabled. Communicate with me.

*G   I require a pilot. (When made by fishing vessels operating in close proximity on the fishing grounds it means 'I am hauling nets'.)

*H   I have a Pilot on board.

*I   I am altering my course to port.

J	I am on fire and have a dangerous cargo on board, keep well clear of me.
K	I wish to communicate with you.
L	You should stop your vessel instantly.
M	My vessel is stopped and making no way throught the water.
N	No (Negative or 'The signification of the previous group should be read in the negative'.) This sign may only be used visually or by sound.
O	Man overboard.
P	In harbour All persons should report on board as the vessel is about to proceed to sea.
P	At Sea. It may be used by fishing vessels to mean 'My nets have come fast upon an obstruction'.
P	I require a Pilot (when made by sound).
Q	My vessel is 'healthy' and I request free pratique.
*S	I am operating astern propulsion.
*T	Keep clear of me. I am engaged in pair trawling.
U	You are running into danger.
V	I require assistance.
W	I require Medical Assistance.
X	Stop carrying out your intentions and watch for my signals.
Y	I am dragging my anchor.
*Z	I require a tug. When made by fishing vessels operating in close proximity on the fishing grounds it means 'I am shooting nets'.

NOTES—1. Signal of letters marked * when made by sound may only be made in compliance with the requirements of the *International Regulations for Preventing Collisions at Sea*.

2. Signals K and S have special meanings as Landing signals for small boats with crews or persons in distress (*International Convention for Safety of Life at Sea*).

Single-letter Signals with Complements

May be made by any method of signalling

A	With three numerals	Azimuth or Bearing
C	With three numerals	Course
D	With two, four or six numerals	Date
G	with four or five numerals	Longitude (the last two numerals denote minutes and the rest degrees).
K	With one numeral	I wish to communicate with you by ... (compliment Table 1).
L	With four numerals	Latitude (the first two denote degrees and the rest minutes).
R	With one or more numerals	Distance in nautical miles.
S	With one or more numerals	Speed in knots.
T	With four numerals	Local Time (the first two denote hours and the rest minutes).
V	With one or more numerals	Speed in kilometres per hour.
Z	With four numerals	G.M.T. (the first two denote hours and the rest minutes.)

Single-letter Signals between Icebreaker and Assisted Vessels

The following Single-letter signals, when made between an icebreaker and assisted vessel, have only the significations given in this Table and are only to be made by Sound, Visual or Radiotelephony signals.

WM Icebreaker support is now commencing. Use special icebreaker support signals and keep continuous watch for Sound, Visual or Radiotelephony signals.

WO Icebreaker support is finished. Proceed to your destination.

Code Letters or Figures	Icebreakers	Assisted Vessel(s)
A ·—	Go ahead (proceed along the ice channel).	I am going ahead (I am proceeding along the ice channel).
G ——·	I am going ahead, follow me.	I am going ahead. I am following you.
J ·———	Do not follow me (proceed along the ice channel).	I will not follow you (I will proceed along the ice channel).
P ·——·	Slow down	I am slowing down.
N —·	Stop your engines	I am stopping my engines
H ····	Reverse your engines.	I am reversing my engines
L ·—··	You should stop your vessel instantly.	I am stopping my vessel.
4 ····—	Stop. I am icebound	Stop. I am icebound
Q ——·—	shorten the distance between vessels.	I am shortening the distance.
B —···	Increase the distance between vessels.	I am increasing the distance.
5 ·····	Attention.	Attention.
Y —·——	Be ready to take (or cast off) the tow line.	I am ready to take (or cast off) the tow line.
* ·———	Stop your headway (given only to a ship in an ice channel ahead of and approaching or going away from icebreaker.	I am stopping headway

*This signal should NOT be made by Radiotelephone.

NOTES—1. The signal K (—·—) by sound or light may be used by an icebreaker to remind ships of their obligation to listen continuously on their radio.

2. If more than one vessel is assisted, the distances between vessels should be as constant as possible; watch speed of your own vessel and vessel ahead. Should speed of your own vessel go down, give attention signal to the following vessel.

3. The use of these signals does not relieve any vessel from complying with the *International Regulations for Preventing Collisions at Sea.*

Single-letter signals which may be used during icebreaking operations:

E	I am altering my course to starboard.
I	I am altering my course to port.
S	I am operating astern propulsion.
M	My vessel is stopped and making no way through the water.

NOTES—1. Signals of letters marked *, when made by Sound, may only be made in compliance with the requirements of the *International Regulations for Preventing Collisions at Sea.*

2. Additional signals for icebreaking support can be found in Code.

GENERAL SECTION
1. Distress Emergency
ABANDON.

AC	I am abandoning my vessel.
AD	I am abandoning my vessel which has suffered a nuclear accident and is a possible source of radiation danger.
AE	I must abandon my vessel.
AF	I do not intend to abandon my vessel.
AF1	Do you intend to abandon your vessel?

ACCIDENT.

AJ	I have had a serious nuclear accident and you should approach with caution.
AK	I have had nuclear accident on board. I am abandoning my

vessel whch has suffered a nuclear accident and is a possible source of radiation danger.............. *AD*
You should steer course ... (or follow me) to reach position of the accident................................*FL*
Position of the accident (or survival craft) is marked by flame or smoke float. *FJ*1
Position of accident (or survival craft) is marked by sea marker dye..................................*FJ*3

DOCTOR

AL I have a doctor on board.

AM Have you a doctor?

AN I need a doctor.
I require a helicopter urgently, with a doctor........*BR*2

INJURED/SICK

A0 Number of injured and/or dead are not yet known.

A01 How many injured?

A02 How many dead?
I require Helicopter urgently to pick up injured/sick person..*BR*3

AT You should send injured/sick person to me.

AIRCRAFT-HELICOPTER

Alight—Landing.

AU I am forced to alight near you (or in position indicated).

AV I am alighting (in position indicated if necessary) to pick up crew of vessel/aircraft.

AW Aircraft should endeavour to alight where flag is waved or light is shown.

AX You should train your searchlight nearly vertical on a cloud, intermittently if possible, and if my aircraft is seen, deflect the beam up wind and on the water to facilitate my landing.

BB	You may alight on my deck.
BB1	You may alight on my deck; I am ready to receive you forward.
BB3	You may alight on my deck; I am ready to receive you aft.

COMMUNICATIONS.

BC	I have established communications with the aircraft in distress on 2182 kHz.
BC1	Can you communicate with aircraft?
BD	I have established communications with aircraft in distress on... kHz.
BE	I have established communications with the aircraft in distress on... MHz.

DITCHED-DISABLED-AFLOAT.

BF	Aircraft is ditched in position indicated and requires immediate assistance. I sighted disabled aircraft in lat... long... at time indicated.................... *DS*
BG	Aircraft is still afloat.

FLYING

BH	I sighted an aircraft at time indicated in lat... long... flying on course....
BJ	I am circling over the area of the accident.
BJ1	An aircraft is circling over the area of accident.
BL	I have engine trouble but am continuing flight.

PARACHUTE.

BM	You should parachute object to windward. Mark it by smoke or light signal.
BM1	I am going to parachute object to windward, marking it by smoke or light signal.
BO	We are going to jump by parachute.

SEARCH-ASSISTANCE.

BP	Aircraft is coming to participate in search. Expected arrive over area of accident at time indicated.
	Search by aircraft/helicopter will be discontinued because of unfavourable conditions....................... *FV*

SPEED

BQ The speed of my aircraft in relation to the surface of the earth is ... (knots or kilometres per hour).

BQ1 What is the speed of your aircraft in relation to the surface of the earth?

HELICOPTER.

BR I require a helicopter urgently.

BS You should send a helicopter/boat with stretcher.

BT Helicopter is coming to you now (or at time indicated).

BV I cannot send a helicopter.

BW The magnetic course for you to steer towards me (or vessel or position indicated) is ... (at time indicated).

ASSISTANCE.

I am in distress and require immediate assistance......*NC*

CB I require immediate assistance.

CB5 I require immediate assistance: I am drifting.

CB8 I require immediate assistance: propeller shaft is broken.

CC I am (or vessel indicated is) in distress in lat ... long ... (or bearing ... from place indicated distance ...) and require immediate assistance (complements Table II, if required).

CD I require assistance in the nature of ... (Complements Table II).

CF Signals from vessel/aircraft requesting assistance are coming from bearing ... from me (lat.... long.... if necessary).

CH Vessel indicated is reported as requiring assistance in lat.... long.... (or bearing ... from place indicated distance ...).

CH1 Light-vessel (or lighthouse) indicated requires assistance.

CH2 Space-ship is down in lat.... long.... and requires immediate assistance.

CI Vessel aground in lat.... long.... requires assistance.

NOT REQUIRED—DECLINED.

CK Assistance is not (or is no longer) required by me (or vessel indicated).

CL	I offered assistance but it was declined.

GIVEN—NOT GIVEN.

CM	One or more vessels are assisting the vessel in distress?
CN	You should give all possible assistance.
CN1	You should give immediate assistance to pick up survivors.
CO	Assistance cannot be given to you (or vessel/aircraft indicated).

PROCEEDING TO ASSISTANCE.

CP	I am (or vessel indicated is) proceeding to your assistance.
CS[1]	What is the name or identity signal of your vessel (or station)?
CU	Assistance will come at time indicated.
CU1	I can assist you.

BOATS—RAFTS.

CW	Boat/raft is on board.
CW1	Boat/raft is safe.
CW7	Boat/raft has sunk.
CW8	Boat/raft has capsized.
CY	Boat(s) is (are) coming to you.
DD	Boats are not allowed to come alongside.
DD1	Boats are not allowed to land (after time indicated).

AVAILABLE.

DF	I have ... (number) serviceable boats.
DH	I have no boat/raft.
DH1	I have no motor boat.

REQUIRED.

DI	I require boats for ... (number) persons.
DJ	Do you require a boat?

SEND.

DK	You should send all available boats/rafts.
DL	I can send a boat.
	I cannot send a boat. *CX*7

---

[1] Procedural signal.

SEARCH

DM   You should search for the boat(s)/raft(s).
DN   I have found the boat/raft.
DP   There is a boat/raft in bearing... distance... from me (or from position indicated).

DISABLED—DRIFTING—SINKING

DISABLED.

DR   Have you sighted disabled vessel/aircraft in approximate lat.... long....?
DT   I sighted disabled vessel in lat.... long.... at time indicated.
DT1   I sighted disabled vessel in lat.... long.... at time indicatd, apparently without a radio.

DRIFTING.

DU   I am drifting at... (number) knots, towards... degrees.
DV   I am drifting.
DV1   I am adrift.
           I require immediate assistance; I am drifting.........*CB*5
           I have broken adrift...........................*RC*1

SINKING.

DX   I am sinking (lat.... long.... if necessary).
DY   Vessel (name or identity) has sunk in lat.... long....
DY1   Did you see vessel sink?
DY2   Where did vessel sink?
DY3   Is it confirmed that vessel (name or identity signal) has sunk?
DY4   What is the depth of water where vessel sunk?

DISTRESS.

DZ   Vessel (or aircraft) indicated appears to be in distress.
DZ1   Is vessel (or aircraft) indicated in distress?
EA   Have you sighted or heard of a vessel in distress (approximate position lat.... long.... or bearing... from place indicated, distance...).

BROWN'S SIGNALLING

EC   A vessel which has suffered a nuclear accident is in distress in lat. ... long. ...

ED   Your distress signals are understood.

ED1   Your distress signals are understood, the nearest lifesaving station is being informed.

EG   Did you hear SOS/MAYDAY given at time indicated?

EG1   Will you listen on 2182kHz for signals of emergency position-indicating Radio beacons?

EK   I have sighted distress signal in lat. ... long. ...

EK1   An explosion was seen or heard (position or direction and time to be indicated).

POSITION OF DISTRESS.

EL   Repeat the distress position.

EL1   What is the position of vessel in distress?
Position given with SOS/MAYDAY from vessel (or aircraft) was lat. ... long. ... (or bearing ... from place indicated, distance ...). .... *FG*
What was the position given with SOS/MAYDAY from vessel (or aircraft)? *FG1*
Position given with SOS/MAYDAY is wrong. The correct position is lat. ... long. ... *FH*

EM   Are there other vessel/aircraft in the vicinity of vessel/aircraft in distress?

CONTACT OR LOCATE.

EN   You should try to contact vessel/aircraft in distress.

EP   I have lost sight of you.
I am flying to likely position of vessel in distress. ...... *BI*
Vessel/aircraft reported in distress is receiving assistance. *CM*1
I have found vessel/aircraft in distress in lat. ... long. ... . *GF*

POSITION.

ER   You should indicate your position at time indicated.

ET   My position at time indicated was lat. ... long. ...

EU   My present position is lat. ... long. ... (or bearing ... from place indicated, distance ...).

EW   My position is ascertained by dead reckoning.

EW1	My position is ascertained by visual bearings.
EW4	My position is ascertained by radar.
EX	My position is doubtful.
EZ	Your position according to bearings taken by radio direction-finder stations which I control is lat.... long.... (at time indicated).
EZ1	Will you give me my position according to bearings taken by radio direction-finder stations which you control?

POSITION OF DISTRESS.

FC	You should indicate your position by visual or sound signals.
FC1	You should indicate your position by rockets or flares.
FC4	You should indicate your position by searchlight.
FD	My position is indicated by visual or sound signals.
FD1	My position is indicated by rockets or flares.
FD2	My position is indicated by visual signals.
FD3	My position is indicated by sound signals.
FD4	My position is indicated by searchlight
	I expect to be at the position of vessel/aircraft in distress at time indicated. ....*EQ*

SEARCH AND RESCUE.

Assistance, Proceeding to

	I am proceeding to the assistance of vessel/aircraft in distress. (lat.... long....)..........................*CR*
FE	I am proceeding to the position of accident at full speed. Expect to arrive at time indicated.
FE1	Are you proceeding to position of accident; if so, when do you expect to arrive?

POSITION OF DISTRESS OR ACCIDENT.

FF	I have intercepted SOS/MAYDAY from vessel (name or identity signal) (or aircraft) in position lat.... long.... at time indicated.

FF1	I have intercepted SOS/MAYDAY from vessel (name or identity signal) (or aircraft) in position lat.... long.... at time indicated; I have heard nothing since.
FJ	Position of accident (or survival craft) is marked.
FJ1	Position of accident (or survival craft) is marked by flame or smoke float.
FJ2	Position of accident (or survival craft) is marked by sea marker.
FJ3	Position of accident (or survival craft) is marked by sea marker dye.
FJ4	Position of accident (or survival craft) is marked by radio beacon.
FJ5	Position of accident (or survival craft) is marked by wreckage.

INFORMATION—INSTRUCTIONS.

FL	You should steer course ... (or follow me) to reach position of accident.
	Course to reach me is............................ *MF*
	What is the course to reach you? *MF*1
FN	I have lost all contact with vessel.
FO	I will keep close to you.
FO1	I will keep close to you during the night.
FQ	You should transmit your identification and series of long dashes or your carrier frequency to home vessel (or aircraft) to your position.
FQ1	Shall I home vessel (or aircraft) to my position?

SEARCH.

FR	I am (or vessel indicated is) in charge of co-ordinating search.
FS	Please take charge of search in sector stretching between bearings ... and ... from vessel in distress.
FW	You should search in the vicinity of lat.... long....
FY	I am in the search area.
FY1	Are you in the search area?
GA	I cannot continue to search.
GB	You should stop search and return to base or continue your voyage.

RESULTS OF SEARCH.

GC	Report results of search.
GC1	Results of search negative. I am continuing to search.
GC2	I have searched area of accident but found no trace of derelict or survivors.
GC3	I have noted patches of oil at likely position of accident.
GD	Vessel/aircraft missing or being looked for has not been heard of since.
GF	I have found vessel/aircraft in distress in lat. . . , long. . . . .
GJ	Wreckage is reported in lat. . . . long. . . . .
GJ1	Wreckage is reported in lat. . . . long. . . . No survivors appear to be in the vicinity.
GL	I have located (or found) wreckage from the vessel/aircraft in distress (position to be indicated if necessry by lat. . . . long. . . . or by bearing . . . from specified place and distance).

RESCUE.

GM	I cannot save my vessel.
GM1	I cannot save my vessel; keep as close as possible.
GN	You should take off persons.
GN1	I wish some persons taken off. Skeleton crew will remain on board.
GQ	I cannot proceed to the rescue owing to the weather. You should do all you can.
GR	Vessel coming to your rescue (or to the rescue of vessel or aircraft indicated) is steering course . . . speed . . . knots.
GU	It is NOT safe to fire a rocket.
GW	Man overboard. Please take action to pick him up (position to be indicated if necessary).
	Man overboard. . . . . *O*

RESULTS OF RESCUE.

GX	Report results of rescue.
GX1	What have you (or rescue vessel/aircraft) picked up?
GY	I (or rescue vessel/aircraft) have picked up wreckage.
GZ	All persons saved.
GZ1	All persons lost.

SURVIVORS.

HF	I have located survivors in water, lat.... long.... (or bearing... from place indicated, distance...).
HJ	I have located survivors on drifting ice, lat.... long....
HL	Survivors not yet located.
HL1	I am still looking for survivors.
HL2	Have you located survivors? If so, in what position?
HM	Survivors are in bad condition. Medical assistance is urgently required.
HM1	Survivors are in bad condition.
HM2	Survivors are in good condition.
HM4	What is the condition of survivors?
HP	Survivors have not yet been picked up.
HP1	Have survivors been picked up?
HR	You should try to obtain from survivors all possible information.

CASUALTIES—DAMAGE.

Collision

HV	Have you been in collision?
HW	I have (or vessel indicated has) collided with surface craft.
HW1	I have (or vessel indicated has) collided with light-vessel.
HW2	I have (or vessel indicated has) collided with submarine
HW3	I have (or vessel indicated has) collided with unknown vessel.
HW4	I have (or vessel indicated has) collided with underwater object.
HX	Have you received any damage in collision?
HX1	I have received serious damage above the waterline.
HX2	I have received serious damage below the waterline.
HY	The vessel (name or identity signal) with which I have been in collision has sunk.
HZ	There has been a collision between vessels indicated (name or identity signal).

I urgently require a collision mat.................. *KA*
I have placed collision mat. I can proceed without assistance..................................... *KA*1

DAMAGE—REPAIRS.

IA	I have received damage to stern
IA1	I have received damage to stern frame.
IA4	I have received damage to bottom plate.
IB	What damage have you received?
IB1	My vessel is seriously damaged.
IB2	I have minor damage.
IC	Can damage be repaired at sea?
ID	Damage can be repaired at sea,
IF	Damage cannot be repaird at sea.
IL	I can only proceed at slow speed.
IL1	I can only proceed with one engine.
IL2	I am unable to proceed under my own power.
IL3	Are you in a condition to proceed?
	Propeller shaft is broken .... *RO*

DIVER—UNDERWATER OPERATIONS.

IN	I require a diver.
IN1	I require a diver to clear propeller.
IO	I have no diver.
IP	A diver will be sent as soon as possible (or at time indicated).
IR	I am engaged in submarine survey work (underwater operations). Keep clear of me and go slow.
	NOTE—The use of this signal does not relieve any vessel from compliance with Rule 18 of the *International Regulations for Preventing Collisions at Sea*.

FIRE—EXPLOSION

FIRE.

IT	I am on fire.
IT1	I am on fire and have a dangerous cargo on board; keep well clear of me.
IT2	Vessel (name or identity signal) is on fire.
IT3	Are you on fire?
IV	Where is the fire?
IV1	I am on fire in the engine-room.

IV2	I am on fire in the boiler-room.
IV3	I am on fire in hold or cargo.
IV4	I am on fire in passengers' or crew's quarters.
IV5	Oil is on fire.
IW	Fire is under control.
IX	Fire is gaining.
IX1	I cannot get the fire under control without assistance.
IZ	Fire has been extinguished.
JA	I require fire-fighting appliances.
JA1	I require foam fire extinguishers.
JA7	I require water pumps.

EXPLOSION.

JB	There is danger of explosion.
JC	There is no danger of explosion.
JC1	Is there any danger of explosion?
JD	Explosion has occurred in boiler.
JD1	Explosion has occurred in tank.
JD2	Explosion has occurred in cargo.
JE	Have you any casualties owing to explosion?

GROUNDING—BEACHING—REFLOATING
GROUNDING

JF	I am (or vessel indicated is) aground in lat.... long.... (also the following complements if necessary):
0	On rocky bottom.
1	On soft bottom.
2	Forward.
3	Amidship.
4	Aft.
5	At high water forward.
6	At high water amidship.
7	At high water aft.
8	Full length of vessel.
9	Full length of vessel at high water.
JG	I am aground; I am in a dangerous situation.

JH	I am aground; I am not in danger. I require immediate assistance; I am aground *CB*4
JJ	My maximum draught when I went aground was...(number) feet or metres.
JL	You are running the risk of going aground.
JM	You are running the risk of going aground at low water.

BEACHING.

JN	You should beach the vessel in lat.... long....
JN1	You should beach the vessel where flag is waved or light is shown.
JN2	I must beach the vessel.

REFLOATING.

JO	I am afloat.
JO1	I am afloat forward.
JO2	I am afloat aft.
JO3	I may be got afloat if prompt assistance is given.
JO4	Are you (or vessel indicated) still afloat?
JO5	When do you expect to be afloat?
JQ	I cannot refloat without jettisoning (the following complements should be used if required):
1	Cargo
2	Bunkers
3	Everything moveable forward.
4	Everything moveable aft.
JV	Will you refloat me to lat.... long.... after refloating?

LEAK.

JW	I have sprung a leak.
JZ	Have you sprung a leak?
JZ1	Can you stop the leak?
JZ2	Is the leak dangerous?
KA	I urgently require a collision mat?
KD	There are...(number) compartments flooded.
KE	The watertight bulkheads are standing up well to the pressure of water.
KE1	I need timber to support bulkheads.

TOWING—TUGS
TUG.

KF	I require a tug (or . . . (number) (tugs).
KI	There are no tugs available.
KI1	Tugs cannot proceed out.

Towing—Taking in Tow.

KJ	I am towing a submerged object.
KJ1	I am towing a float.
KJ2	I am towing a target.
KL	I am obliged to stop towing temporarily.
KL1	You should stop towing temporarily.
KN	I cannot take you (or vessel indicated) in tow.
KP	You should tow me to the nearest port or anchorage (or place indicated).
KQ	Prepare to be taken in tow.
KQ1	I am ready to be taken in tow.
KQ2	Prepare to tow me (or vessel indicated).
KQ3	I am ready to tow you.
KR	All is ready for towing.
KR1	I am commencing to tow.
KR2	You should commence towing.
KR3	Is all ready for towing?

TOWING LINE.
Cable Hawser.

KS	You should send a line over.
KS1	I have taken the line.
KT	You should send me a towing hawser.
KT1	I am sending towing hawser.
KU	I cannot send towing hawser.
KU1	I have no, or no other, hawser.
KU2	I have no wire hawser.
KU3	Have you a hawser?
KW	You should have towing hawser/cable ready.
KW1	Towing hawser/cable is ready.
KX	You should be ready to receive the towing hawser.

KX1	I am ready to receive the towing hawser.
KY	Length of tow is... (number) fathoms.
LA	Towing hawser/cable has parted.

MAKE FAST.

Veer.

LB	You should make towing hawser fast to your chain cable.
LB1	Towing hawser is fast to chain cable.
LC	You should make fast astern and steer me.
LD	You should veer your hawser/cable... (number) fathoms.
LE	I am about to veer my hawser/cable.
LF	You should stop veering your hawser/cable.
LF1	I cannot veer any more hawser/cable.

CAST OFF.

LG	You should prepare to cast off towing hawser(s).
LG1	I am preparing to cast off towing hawser(s).
LG2	I am read to cast off towing hawser(s).
LG3	You should cast off starboard towing hawser.
LG4	I have cast off starboard towing hawser.
LG5	You should cast off port towing hawser.
LG6	I have cast off port towing hawser.

ENGINES.

Manoeuvres.

I am going ahead.	QD
My engines are going ahead.	QD1
I will keep going ahead.	QD2
I have headway.	QE
I cannot go ahead.	QF
You should go slow ahead.	QG1
I am going astern.	QI
My engines are going astern.	QI1
I cannot go astern.	QK
You should not go astern any more.	QM
You should stop your engines immediately.	RL
You should stop your engines.	RL1

BROWN'S SIGNALLING

	My engines are stopped.............................. *RM*
LI	I am increasing speed.
LI1	Increase speed.
LJ	I am reducing speed.
LJ1	Reduce speed.

AIDS TO NAVIGATION—NAVIGATION—HYDROGRAPHY
AIDS TO NAVIGATION

BUOYS—BEACONS.

LK	Buoy (or beacon) has been established in lat... long....
LL	Buoy (or beacon) in lat.... long.... has been removed.
	You should steer directly for the buoy (or object indicated)......................................*PL*
	You should keep buoy (or object indicated) on your starboard side. *PL*1
LM	Radio Beacon indicated is out of action.

LIGHTS—LIGHT-VESSELS.

LN	Light (name follows) has been extinguished.
LN1	All lights are out along this coast (or coast of...).
LO	I am not in my correct position (to be used by a light-vessel).
LO1	Light-vessel (name follows) is out of position.
LO2	Light-vessel (name follows) has been removed from her station.
	Light-vessel (or lighthouse) indicated requires assistance....................................... *CH*1

BAR

LP	There is not less than... (number) feet or metres over the bar.
LQ	There will be... (number) feet or metres of water over the bar at time indicated.
LR	Bar is not dangerous.
LR1	What is the depth of water over the bar?
LR2	Can I cross the bar?
LS	Bar is dangerous.

BEARINGS.

LT	Your bearing from me (or from ...) (name or identity signal) is ... (at time indicated).
LV	Let me know my bearings from you. I will flash searchlight.
LV1	What is my bearing from you [or from ... (name or identity signal)]?
LV2	What is the bearing of ... (name or identity signal) from ... (name or identity signal)?
	Your magnetic bearing from me (or from vessel or position indicated) is ... (at time indicated) .... *BZ*
	What is my magnetic bearing from you (or from vesel or position indicated)? .... *CA*
LW	I receive your transmission on bearing ...
	Your position according to bearings taken by radio direction-finder stations which I control is lat.... long.... (at time indicated). .... *EZ*
	What is the bearing and distance by radar of vessel (or object) indicated? *OM*1

CANAL—CHANNEL—FAIRWAY

CANAL.

LX	The canal is clear.
LX1	The canal will be clear at time indicated.
LX2	You can enter the canal at time indicated.
LX3	Is the canal clear?
LX4	When can I enter the canal?
LY	The canal is not clear.
LZ	The channel/fairway is navigable.
LZ1	I intend to pass through the channel/fairway.
LZ2	Is the channel/fairway navigable?
MA	The least depth of water in the channel/fairway is ... (number of feet or metres).
MB	You should keep in the centre of the channel/fairway.
MB1	You should keep on the starboard side of the channel/fairway.
MB2	You should keep on the port side of the channel/fairway.
MB3	You should leave the channel/fairway free.
MC	There is an uncharted obstruction in the channel/fairway. You should proceed with caution.

MC1 The channel/fairway is not navigable.

COURSE.

MD My course is...

MD1 What is your course?
My present position, course and speed are lat...., long...., course.... knots... .... *EV*
What are your present position, course and speed?... *EV*1

ME The course to place (name follows) is...

ME1 What is the course to place (name follows)?

MF Course to reach me is...

MF1 What is the course to reach you?
You should maintain your present course. .... *PI*
I cannot maintain my present course. .... *PJ*

MH You should alter course to... (at time indicated).

MI I am altering course to...
I am altering my course to starboard. .... *E*
I am altering my course to port...................... *I*

DANGERS TO NAVIGATION—WARNINGS
DERELICT—WRECK—SHOAL.

MJ Derelict dangerous to navgation reported in lat.... long.... (or Complements Table III).

MK I have seen derelict (in lat.... long.... at time indicated).

MM There is a wreck in lat.... long....

MM1 Wreck is buoyed.

MM2 Wreck is awash.

MN Wreck (in lat.... long....) is not buoyed.

MO I have struck a shoal or submerged object (lat.... long....).

MP I am in shallow water. Please direct me how to navigate.

RADIATION DANGER.

MQ There is a risk of contamination due to excessive release of radioactive material in this area (or area around lat.... long....) Keep radio watch. Relay the message to all vessels in your vicinity.

MQ1	The radioactive material is airborne.
MQ2	The radioactive material is waterborne.
MR	There is no, or no more, risk of contamination due to excessive release of radioactive material in this area (or in area around lat. . . . long. . . .).
MR1	Is there risk of contamination due to excessive release of radioactive material in this area (or in area around lat. . . . long. . . .)?
MS	My vessel is a dangerous source of radiation.
MT	My vessel is a dangerous source of radiation You may approach from . . . (Complements Table III).
MU	My vessel is a dangerous source of radiation. Do not approach within . . . (number) cables.
MW	My vessel is releasing radioactive material and presents a hazard. Do not approach within . . . (number) cables.
MX	The radioactive material is airborne. Do not approach from leeward.

WARNINGS.

MY	It is dangerous to stop.
MY1	It is dangerous to remain in present position.
MY2	It is dangerous to proceed on present course.
MY3	It is dangerous to proceed until weather permits.
MY4	It is dangerous to alter course to starboard.
MY5	It is dangerous to alter course to port.
	It is not safe to fire a rocket. . . . . *GU*
MZ	Navigation is dangerous in the area around lat. . . . long. . . . .
NA	Navigation is closed.
NA1	Navgation is possible only with tug assistance.
NA2	Navigation is possible only with pilot assistance.
	There is fishing-gear in the direction you are heading (or in direction indicated—Complements Table).
NC	I am in distress and require immediate assistance.
ND	Tsunami (phenomenal wave) is expected. You should take appropriate precautions.
NE	You should proceed with great caution.
NF	You are running into danger.

NH	You are clear of all danger.
NH1	Are you clear of all danger?
NI	I have (or vessel indicated has) a list of ... (number) degrees to starboard.
NJ	I have (or vessel indicated has) a list of .... (number) degrees to port.

DEPTH—DRAUGHT

DEPTH.

NK	There is not sufficient depth of water.
NL	There is sufficient depth of water.
NL1	Is there sufficient depth of water?
	What is the least depth of water in the channel/fairway?..................................*LZ*4
	There will be ... (number feet or metres) of water over the bar at time indicated............................*LQ*
	What is the depth at high and low water here (or in place indicated)?................................. *PW*2
NM	You should report the depth around your vessel.
NO	Negative—'No' or 'The significance of the previous group should be read in the negative', (procedural signal).
NP	The depth of water at the bow is ... (number feet or metres).
NQ	The depth of water at the stern is ... (number feet or metres).
NR	The depth of water along the starboard side is ... (number of feet or metres).
NS	The depth of water along the port side is ... (number feet or metres).

DRAUGHT.

NT	What is your draught?
NT1	What is your light draught?
NT2	What is your ballast draught?
NT3	What is your loaded draught?
NT4	What is your summer draught?
NU	My draught is ... (number feet or metres).
NV	My light draught is ... (number feet or metres).
NW	My ballast draught is ... (number feet or metres).

	My loaded draught is...(number feet or metres).
NY	My summer draught is...(number feet or metres).
OE	Your draught must not exceed...(number feet or metres).
OF	I could lighten to...(number feet or metres draught).
OG	To what draught could you lighten?

ELECTRONIC NAVIGATION

RADAR.

OH	You should switch on your radar and keep radar watch.
OH1	The restrictions on the use of radar are lifted.
OI	I have no radar.
OI1	Are you equipped with radar?
OI2	Is your radar in operation?
OJ	I have located you on my radar bearing..., distance...miles.
OJ1	I cannot locate you on my radar.
OL	Is radar pilotage being effected in this port (or port indicated).
OM	Bearing and distance by radar of vessel (or object) indicated, is bearing...distance...miles.
ON	I have no echo on my radar bearing...distance...miles.

RADIO DIRECTION-FINDER

OO	My radio direction-finder is inoperative.
OP	I have requested...(name or identity signal) to send two dashes of ten seconds each or the carrier of his transmitter followed by his call sign.
OP1	Will you request...(name or identity signal) to send two dashes of ten seconds each or the carrier of his transmitter followed by his call sign?
OP2	Will you send two dashes of ten seconds each, or the carrier of your transmitter, followed by your call sign?
OQ	I am calibrating radio direction-finder or adjusting compasses.

DECCA—LORAN—CONSOL.

My position is ascertained by Decca Navigator..... *EW*5

My position is ascertained by Loran *EW*6

My position is ascertained by Consol *EW7*

MINES—MINESWEEPING.

OR	I have struck a mine.
	I have a mine in my sweep (or net) .... *TO*
OS	There is a danger from mines in this area (or area indicated).
OS1	You should keep a look-out for mines.
OS2	You are out of the dangerous zone.
OS3	Am I out of the dangerous zone?
OS4	Are you out of the dangerous zone?
OS5	Is there any danger from mines in this area (or area indicated)?
OT	Mine has been sighted in lat.... long.... (or in direction indicated Complements Table III).
OV	Mine(s) is (are) believed to be bearing... from me, distance... miles.
OW	There is a Minefield ahead of you. You should stop your vessel and wait instructions.
OW1	There is a minefield along the coast. You should not approach too close.
OX	The approximate direction of the minefield is bearing... from me.
OY	Port is mined.
OY1	Entrance is mined.
OY2	Fairway is mined.
OY3	Are there mines in the port, entrance or fairway?
OZ	The width of the swept channel is... (number feet or metres).
PA	I will indicate the swept channel. You should follow in my wake.
PA1	You should keep carefully to the swept channel.
PA2	The swept channel is marked by buoys.
PA3	I do not see the buoys marking the swept channel.
PA4	Do you know the swept channel?
PB	You should keep clear of me. I am engaged in minesweeping operations.
PB1	You should keep clear of me. I am exploding a floating mine.

PC	I have destroyed the drifting mine(s).
PC1	I cannot destroy the drifting mine(s).

NAVIGATION LIGHTS—SEARCHLIGHT.

PD	Your navigation light(s) is (are) not visible.
PD1	My navigation lights are not functioning.
PE	You should extinguish all the lights except the navigation lights.
PG	I do not see any lights.
PG1	You should hoist a light.
PG2	I am dazzled by your searchlight. Douse it or lift it.
	You should train your searchlight nearly vertical on a cloud, intermittently if possible, and if my aircraft are seen, deflect the beam up wind and on the water to facilitate my landing............................ *AX*
	Shall I train my searchlight nearly vertical on a cloud, intermittently if possible, and, if your aircraft is seen, deflect the beam up wind and on the water to facilitate your landing?....................................... *AX*1

NAVIGATING AND STEERING INSTRUCTIONS

PH	You should steer as indicated.
PH1	You should steer towards me.
PH2	I am steering towards you.
PH3	You should steer more to starboard.
PH4	I am steering more to starboard.
PH5	You should steer more to port.
PH6	I am steering more to port.
PI	You should maintain your present course.
PI1	I am maintaining my present course.
PI2	Shall I maintain my present course?
PK	I cannot steer without assistance.
PL	You should steer directly for the buoy (or object indicated).
PL1	You should keep buoy (or object indicated) on your starboard side.
PL2	You should keep buoy (or object indicated) on your port side.
PL3	You can pass the buoy (or object indicated) on either side.

PN	You should keep to leeward of me (or vessel indicated).
PN1	You should keep to windward of me (or vessel indicated).
PN2	You should keep on my starboard side (or starboard side of vessel indicated).
PN3	You should keep on my port side (or port side of vessel indicated).
PO	You should pass ahead of me. (or vessel indicated)
PO1	I will pass ahead of you (or vessel indicated).
PO2	You should pass astern of me (or vessel indicated).
PO3	I will pass astern of you (or vessel indicated).
PP	Keep well clear of me.
PP1	Do not overtake me.
PP2	Do not pass ahead of me.
PP3	Do not pass astern of me.
PS	You should not come any closer.
PS1	You should keep away from me (or vessel indicated).
	I am calibrating radio direction-finder or adjusting compasses. .... *OQ*

TIDE.

PT	What is the state of the tide?
PT1	The tide is rising.
PT2	The tide is falling.
PT3	The tide is slack.
PU	The tide begins to rise at time indicated.
PU1	When does the tide begin to rise?
PV	The tide begins to fall at time indicated.
PV1	When does the tide begin to fall?
PW	What is the rise and fall of the tide?
PW1	What is the set and drift of the tide?
PW2	What is the depth at high and low water here (or in place indicated)?
PX	The rise and fall of the tide is . . . (number feet or metres).
PY	The set of the tide is . . . degrees.
PZ	The drift of the tide is . . . knots.
QA	The depth at high water here (or in place indicated) is . . . (number feet or metres).

QB	The depth at low water here (or in place indicated) is ... (number feet or metres).
QC	You should wait until high water.
QC1	You should wait until low water.

MANOEUVRES—AHEAD—ASTERN
AHEAD—HEADWAY.

QD	I am going ahead.
QD1	My engines are going ahead.
QD2	I will keep going ahead.
QD3	I will go ahead.
QD4	I will go ahead dead slow.
QE	I have headway.
QF	I cannot go ahead.
QG	You should go ahead.
QG1	You should go slow ahead.
QG2	You should go full speed ahead.
QG3	You should keep going ahead.
QG4	You should keep your engines going ahead.
QH	You should not go ahead any more.

ASTERN—STERNWAY.

QI	I am going astern.
QI1	My engines are going astern.
QI2	I will keep going astern.
QJ	I have sternway.
QK	I cannot go astern.
QL	You should go astern.
QL1	You should go slow astern.
QL2	You should go full speed astern.
QL3	You should keep going astern.
QL4	You should keep your engines going astern.
QM	You should not go astern any more.

ALONGSIDE.

QN	You should come alongside my starboard side.
QN1	You should come alongside my port side.

QN2	You should drop an anchor before coming alongside.
QO	You should not come alongside.
QP	I will come alongside.
QP1	I will try to come alongside.
QQ	I require health clearance.
QR	I cannot come alongside.
QR1	Can I come alongside?

TO ANCHOR—ANCHOR(S)—ANCHORAGE

TO ANCHOR.

QS	You should anchor at time indicated.
QS1	You should anchor (position to be indicated if necessary).
QS2	You should anchor to await tug.
QS3	You should anchor with both anchors.
QS4	You should anchor as convenient.
QS5	Are you going to anchor?
QT	You should not anchor. You are going to foul my anchor.
QU	Anchoring is prohibited.
QV	I am anchoring in position indicated.
QV1	I have anchored with both anchors.
QW	I shall not anchor.
QW1	I cannot anchor.
QX	I request permission to anchor.
QX1	You have permission to anchor.
QY	I wish to anchor at once.
QY1	Where shall I anchor?

ANCHOR (S).

QZ	You should have your anchors ready for letting go.
QZ1	You should let go another anchor.
RA	My anchor is foul.
RA1	I have picked up telegraph cable with my anchor.
RB	I am dragging my anchor.
RB1	You appear to be draggin your anchor.
RB2	Where you have anchored (or intend to anchor) you are likely to drag.

RC	I am (or vessel indicated is) breaking adrift.
RC1	I have broken adrift.
RD	You should weigh (cut or slip) anchor immediately.
RD1	You should weigh anchor at time indicated.
RD2	I am unable to weigh my anchor.

ANCHORAGE.

RE	You should change your anchorage/berth. It is not safe.
RF	Will you lead me into safe anchorage?
	You should tow me to nearest port or anchorage (or place indicated) ....*KP*
	I will tow you to nearest port or anchorage (or place indicated) *KP*1
	I must get shelter or anchorage as soon as possible ..*KP*2
RG	You should send a boat to where I am to anchor or moor.
RG1	At what time shall I come into anchorage?
	You should proceed to anchorage in position indicated (lat.... long....). ....*RW*
	You should not proceed out of harbour/anchorage ..*RZ*1
RH	There is no good holding ground in my area (or around lat.... long....).
RI	There is good holding ground in my area (or around lat.... long....).
RI1	Is there good holding ground in your area. (or around lat.... long....)?

ENGINES—PROPELLER

ENGINES.

RJ	You should keep your engines ready.
RJ1	You should have your engines ready as quickly as possible.
RJ2	You should report when your engines are ready.
RJ3	You should leave when your engines are ready.
RJ4	At what time will your engines be ready?
RK	My engines will be ready at time indicated.
RK1	My engines are ready.
RL	You should stop your engines immediately.
RL1	You should stop your engines.
RM	My engines are stopped.

RM1	I am stopping my engines.
RM2	I am obliged to stop my engines.
RN	My engines are out of action.
	I can only proceed with one engine *IL*1

PROPELLER.

RO	Propeller shaft is broken.
RO1	My propeller is fouled by hawser or rope.
RO2	I have lost my propeller.
	I require immediate assistance; propeller shaft is broken..*CB*8

LANDING—BOARDING

LANDING.

RP	Landing here is highly dangerous.
RP1	Landing here is highly dangerous. A more favourable location for landing is at position indicated.
RQ	Interrogative or 'the signification of the previous group should be read as a question'. (Procedural signal.)
RR	This is the best place to land.
RR1	Lights will be shown or flag waved at the best landing place.
	Boat should endeavour to land where flag is waved or light is shown .... *DC*
	Boats are not allowed to land after (time indicated). *DD*1

BOARDING.

RS	No one is allowed on board.
	You should stop or heave to, I am going to board you. *SQ*3

MANOEUVRES.

RT **X**	Stop carrying out your intentions and watch for my signals.
RT1	What manoeuvres do you intend to carry out?
RU	Keep clear of me; I am manoeuvring with difficulty. ... *D*
RU1	I am carrying out manoeuvring trials.

PROCEED—UNDER WAY

PROCEED.

RV	You should proceed to (place indicated if necessary).

RV1	You should proceed to destination.
RV2	You should proceed into port.
RV3	You should proceed to sea.
RW	You should proceed to anchorage in position indicated (lat.... long....).
RX	You should proceed at time indicated.
RY	You should proceed at slow speed when passing me (or vessels making this signal).

You should proceed to rescue of vessel)or ditched aircraft) in lat.... long.................................*GP*
You should proceed to lat.... long.... to pick up survivors... *HN*
You should proceed with great caution. ....*NE*
You should proceed with great caution; the coast is dangerous..*NE1*
You should proceed with great caution. Submarines are exercising in this area. *NE2*
You should proceed with great caution: There is a boom across...*NE3*
You should proceed with great caution; keep clear of firing range *NE4*
You should proceed with great caution; hostile vessel sighted (lat.... long....). *NE5*
You should proceed with great caution; hostile submarine sighted (lat.... long....). *NE6*
You should proceed with great caution; hostile aircraft sighted (lat.... long....). *NE7*

RZ	You should not proceed to (place indicated if necessary).
RZ1	You should not proceed out of harbour/anchorage.

All vessels should proceed to sea as soon as possible owing to danger in port...............................*UL*

SA	I can proceed at time indicated.

I am proceeding to position of accident at full speed. Expect to arrive at time indicated.................*FE*
Are you proceeding to position of accident? If so, when do you expect to arrive?..........................*FE1*
I can proceed at ... (number) knots...*IK*
I can only proceed with one engine *IL1*
I am unable to proceed under my own power.*IL2*
Are you in a condition to proceed? *IL3*

UNDER WAY.

SC	I am under way.
SC1	I am ready to get under way.
SC2	I shall get under way as soon as weather permits.
SD	I am not ready to get under way
SF	Are you (or vessel indicated) under way?
SF1	Are you ready to get under way?
SF2	At what time will you be under way?
SG	My present speed is ... (number) knots.
SJ	My maximum speed is ... (number) knots.
SL	What is your present speed?
SL1	What is your maximum speed?

The speed of my aircraft in relation to the surface of the earth is ... (knots or kilometres per hour). . . . . *BQ*

What is the speed of your aircraft in relation to the surface of the earth? *BQ*1

My present position, course and speed are lat. ... long. ... course. ... knots . . . . *EV*

What are your present position course and speed? . . . *EV*1

I can only proceed at slow speed. *IL*

I am increasing speed. . . . . *LI*

Increase speed *LI*1

I am reducing speed . . . . *LJ*

SM	I am undergoing speed trials.

STOP—HEAVE TO.

SN	You should stop immediately. Do not scuttle. Do not lower boats. Do not use wireless. If you disobey I shall open fire.
SO	You should stop your vessel instantly.
SO1	You should stop. Head off shore.
SO2	You should remain where you are.
SP	Take the way off your vessel.
SP1	My vessel is stopped and making no way through the water *M*
SQ	You should stop, or heave to.
SQ1	You should stop or heave to, otherwise I shall open fire on you.
SQ2	You should stop or heave to; I am going to send a boat.
SQ3	You should stop or heave to; I am going to board you.

You should heave to or anchor until pilot arrives. . . . . *UB*

I am (or vessel indicated is) stopped in thick fog. . . . . . *XP*

MISCELLANEOUS

CARGO—BALLAST.

ST	What is your cargo?
SU	My cargo is agricultural products.
SU1	My cargo is coal.
SU2	My cargo is dairy products.
SU3	My cargo is fruit products.
SU4	My cargo is heavy equipment/machinery.
SU5	My cargo is livestock.
SU6	My cargo is lumber.
SU7	My cargo is oil/petroleum products.
SU8	I have a general cargo.
SV	I am not seaworthy due to shifting of cargo or ballast.
SW B	I am taking in, or discharging, or carrying dangerous goods.
SX	You should not discharge oil or oily mixture.
SY	The discharge of oil or oily mixture in this area is prohibited within...(number) miles from the nearest land.

CREW—PERSONS ON BOARD.

SZ	Total number of persons on board is...
TA	I have left...(number) men on board.
TB	...(number) persons have died.
TC	...(number) persons are sick.

I am alighting (in position indicated if necessary) to pick up crew of vessel/aircraft.............................*AV*
    I cannot alight but I can lift crew.................*AZ*
You cannot alight on the deck. Can you lift crew?...*BA*1
All persons saved.................................*GZ*
All persons lost..................................*GZ*1
I (or rescue vessel/aircraft) have rescued...(number) injured persons. .... *HA*
Can I transfer rescued persons to you? .... *HD*

FISHERY.

TD	I am a fishcatch carrier boat.
TD1	I am a mothership for fishing vessel(s).
TD2	Are you a fishing vessel?
TE	I am bottom trawling.
TE1	I am trawling with a floating trawl.
TE2	I am long-line fishing.
TE3	I am fishing with towing lines.
TE4	I am engaged on two-boat fishing operation.
TE5	I am drifting on my nets.
TE6	In what type of fishing are you engaged?
TF	I am shooting purse seine.
TG	I am hauling purse seine.
TH	You should navigate with caution. Small fishing boats are within . . . (number) miles of me.
TI	You should navigate with caution. You are drifting towards my set of nets.
TJ	You should navigate with caution. There are nets with a buoy in this area.
TK	Is there fishing gear set up on my course?
TL	My gear is close to the surface in a direction . . . (Complements Table III) for a distance of . . . miles.
TO	I have a mine in my sweep (net).
TP	Fishing gear has fouled my propeller.
TQ	You have caught my fishing gear.
TS	You should take the following action with your warps:
1	Veer the port (stern) warp.
2	Veer the starboard (fore) warp.
3	Veer both warps.
4	Stop veering.
TU	I have to cut the warps. The trawls are entangled.
TV	Fishing in this area is prohibited.
TV1	Trawling in this area is dangerous because there is an obstruction.
TW	Attention. You are in the vicinity of prohibited fishery limits.
TX	A fishery protection (or fishery assistance) vessel is in lat. . . . long. . . .

TY	I request assistance from fishery protection (or fishery assistance) vessel.
TZ	Can you offfer assistance (Complements Table II)?

PILOT.

UA	Pilot will arrive at time indicated.
UB	You should heave to or anchor until pilot arrives.
	I have a Pilot on board .... *H*
UC	Is a pilot available in this place (or place indicated)?
	I require a pilot .... *G*
UE	Where can I get a pilot (for destination indicated if necessary)?
UF	You should follow pilot boat (or vessel indicated).
	You should follow in my wake (or wake of vessel indicated)...................................... *PM*
	You should go ahead and lead the course.......... *PM*1
UH	Can you lead me into port?
UI	Sea is too rough, pilot boat cannot get off to you.
UJ	Make a starboard lee for the pilot boat.
UJ1	Make a port lee for the pilot boat.
UK	Pilot boat is most likely on a bearing...from you.
UK1	Have you seen the pilot boat?
	Is radar pilotage being effected in this port (or port indicated)?..................................... *OL*

PORT—HARBOUR.

UL	All vessels should proceed to sea as soon as possible owing to danger in port.
UM	The harbour (or port indicated) is closed to traffic.
	You should not proceed out of harbour/anchorage...*RZ*1
UN	You may enter harbour immediately (or at a time indicated).
UO	You must not enter harbour.
UP	Permission to enter harbour is urgently requested. I have an emergency case.
UQ	You should wait outside the harbour (or river mouth).
UQ1	You should wait outside the harbour until daylight.
UR	My estimated time of arrival (at place indicated) is (time indicated).
UR1	What is the estimated time of arrival (at place indicated)?

MISCELLANEOUS.

US	Nothing can be done until time indicated.
US1	Nothing can be done until daylight.
US2	Nothing can be done until tide has risen.
US3	Nothing can be done until visibility improves.
US4	Nothing can be done until weather moderates.
US5	Nothing can be done until draught is lightened.
US6	Nothing can be done until tugs have arrived.
UT	Where are you bound for?
UT1	Where are you coming from?
UU	I am bound for...
UV	I am coming from...
UW	I wish you a pleasant voyage.
UW1	Thank you very much for your co-operation. I wish you a pleasant voyage.
UX	No information available.
	I am unable to answer your question.............. *YK*

EXERCISES.

UY	I am carrying out exercises. Please keep clear of me.

BUNKERS.

UZ	I have bunkers for...(number) hours.
VB	Have you sufficient bunkers to reach port?
VC	Where is the nearest place at which fuel oil is available?
VC1	Where is the nearest place at which diesel oil is available?
VC2	Where is the nearest place at which coal is available?
VD	Bunkers are available at place indicated (or lat.... long....).

FUMIGATION.

VE	I am fumigating my vessel.
	No-one is allowed on board..................... *RS*

IDENTIFICATION.

	What is the name or identity signal of your vessel (or station)?................................. *CS*
VF	You should hoist your identity signal.

METEOROLOGY—WEATHER

CLOUDS.

VG	The coverage of low clouds is ... (number of octants or eights of sky covered).
VH	The estimated height of base of low clouds in hundreds of metres is ...
VI	What is the coverage of low clouds in octants (eights of sky covered)?
VI1	What is the estimated height of base of low clouds in hundreds of metres?

GALE—STORM—TROPICAL STORM

GALE.

VJ	Gale (wind force Beaufort 8-9) is expected from direction indicated (Complements Table III).

STORM.

VK	Storm (wind force Beaufort 10 or above) is expected from direction indicated (Complements Table III).

TROPICAL STORM.

VL	Tropical storm (cyclone, hurricane, typhoon) is approaching. You should take appropriate precautions.
VM	Tropical storm is centred at ... (time indicated) in lat. ... long. ... on course ... speed ... knots.
VN	Have you latest information of the tropical storm (near lat. ... long. ... if necessary)?
	Very deep depression is approaching from direction indicated (Complements Table III). .... *WT*
	There are indications or an intense depression forming in lat. ... long. ... *WU*

ICE.

VO	Have you encountered ice?
VP	What is the character of ice, its development and the effects on navigation?
VQ	Character of ice.
VQ0	No ice.
VQ1	New ice (ice crystals, slush or sludge, pancake ice or ice rind).

VQ2	Young, fast ice (5-15 cm. thick or rotten fast ice).
VQ3	Open drift ice (not more than 5/8 of the water surface is covered by ice-floes).
VQ4	A compressed accumulation of sludge (a compressed mass of sludge or pancake ice, the ice cannot spread).
VR	Ice development:
VR0	No change.
VR1	Ice situation has improved.
VR2	Ice situation has deteriorated.
VR3	Ice has broken up.
VR4	Ice has opened or drifted away.
	No information available .... *UX*
VS	Effects of ice on navigation:
VS0	Unobstructed.
VS1	Unobstructed for power driven vessels built of iron or steel, dangerous for wooden vessels without ice protection.
VS2	Difficult for low powered vessels without the assistance of an ice breaker, dangerous for vessels of weak construction.
VT	Danger of ice accretion on superstructure (for example black frost).
VT1	I am experiencing heavy icing on superstructure.
VU	I have seen ice-field in lat.... long....
VV	Ice patrol ship is not on station.
VV1	Ice patrol ship is on station.

ICEBERGS.

VW	I have seen icebergs in lat.... long....
VX	I have encountered one or more icebergs or growlers (with or without position and time)
VY	One or more iceberg or growlers have been reported (with or without position and time).
VZ	Navigation is dangerous in the area around lat.... long.... owing to iceberg(s).
VZ1	Navigation is dangerous in the area around lat.... long.... owing to floating ice.
VZ2	Navigation is dangerous in the area around lat.... long.... owing to pack ice.

ICE-BREAKER.

WC	I am (or vessel indicated is) fast on ice and require(s) ice-breaker assistance.
WC1	Ice-breaker is being sent to your assistance.
	I require assistance in the nature of ice-breaker. *CD*9
WD	Ice-breaker is not available.
WD1	Ice-breaker cannot render assistance at present.
WE	Navigation channel is being kept open by ice-breaker.
WF	I can give ice-breaker support only up to lat. . . . long. . . .
WH	I can only assist if you will make all efforts to follow.
WI	At what time will you follow at full speed?
WJ	The convoy will start at time indicated from here (or from lat. . . . long. . . .).
WK	You (or vessel indicated) will be number . . . in convoy.
WL	Ice-breaker is stopping work during darkness.
WN	Ice-breaker is stopping work for . . . (number) hours or until more favourable conditions arise.
	You should go astern. . . . . *QL*

ATMOSPHERIC PRESSURE—TEMPERATURE
ATMOSPHERIC PRESSURE.

WP	Barometer is steady.
WP1	Barometer is falling rapidly.
WP2	Barometer is rising rapidly.
WQ	The barometer has fallen . . . (number) millibars during the past three hours.
WR	The barometer has risen . . . (number) millibars during the past three hours.
WS	Corrected atmospheric pressure at sea level is . . . (number) millibars.
WS1	State corrected atmospheric pressure at sea level in millibars.
WT	Very deep depression is approaching from direction indicated (Complements Table III).
WU	There are indications of an intense depression forming in lat. . . . long. . . .

TEMPERATURE.

WV	The air temperature is sub-zero (centigrade).
WV1	The air temperature is expected to be sub-zero (centigrade).

SEA-SWELL

SEA.

WW What are the sea conditions in your area (or around lat.... long....)?

WX The true direction of the sea in tens of degrees is...(number following indicates tens of degrees).

WY The state of the sea is...(Complements 0-9 corresponding to the following Table):

		Height	
		In Metres	In Feet
WY 0	Calm (glassy)	0	0
1	Calm (rippled)	0—0.1	0—$1/3$
2	Smooth (wavelets)	0.1—0.5	$1/3$—$1^2/_3$
3	Slight	0.5—1.25	$1^2/_3$—4
4	Moderate	1.25—2.5	4—8
5	Rough	2.5—4	8—13
6	Very rough	4—6	13—20
7	High	6—9	20—30
8	Very High	9—14	30—45
9	Phenomenal	over 14	over 45

XB The state of the sea is expected to be...(Complements 0—9 as in Table above).

SWELL.

XC What are the swell conditions in your area (or area around lat.... long....)?

XD The true direction of the swell in tens of degrees is...(number following indicates tens of degrees).

XE The state of the swell is...(Complements 0—9 corresponding to the folllowing Table):

0	No swell	
1	Short or middle	Weak—Approximate height 2 m. (6 ft.)
2	Long	
3	Short	
4	Middle	Moderate—Approximate height 2—4 m. (6-12 ft.)
5	Long	
6	Short	
7	Middle	High—Approximate height 4 m. (12 ft.)
8	Long	
9	Confused	

XH The state of the swell is expected to be...(Complements 0-9 as in the Table above).

VISIBILITY—FOG

XI	Indicate visibility.
XJ	Visibility is . . . (number) tenths of nautical miles.
XK	Visibility is variable between . . . and . . . (maximum and minimum in tenths of nautical miles).
XL	Visibility is decreasing.
XL1	Visibility is increasing.
XL2	Visibility is variable.
XM	What is the forecast visibility in my area (or area around lat. . . . long. . . .)?
XN	Visibility is expected to be . . . . (number) tenths of nautical miles.
XP	I am (or vessel indicated is) stopped in thick fog.
XP1	I am entering zone of restricted visibility.

WEATHER—WEATHER FORECAST

XQ	What weather are you experiencing?
XR	Weather is good.
XR1	Weather is bad.
XR2	Weather is moderating.
XR3	Weather is deteriorating.
XS	Weather report is not available.
XT	Weather expected is bad.
XT1	Weather expected is good.
XT2	No change is expected in the weather.
XT3	What weather is expected?
XU	You should wait until weather moderates.
XV	Please give weather forecast for my area (or area around lat. . . . long. . . .) in MAFOR Code.

WIND.

XW	What is the true direction and force of the wind in your area (or area around lat. . . . long. . . .).
XX	True direction of the wind is . . . (Complements Table III).
XY	Wind force in Beaufort Scale . . . (numerals 0-12).
XZ	What is the wind doing?
XZ1	The wind is backing.

XZ2	The wind is veering.
XZ3	The wind is increasing.
XZ4	The wind is squally.
XZ5	The wind is steady in force.
XZ6	The wind is moderating.
YA	What wind direction and force is expected in my area (or area around lat.... long....)?
YB	True direction of the wind is expected to be... (Complements Table III).
YC	Wind force expected in Beaufort Scale... (numerals 0-12).
YD	What is the wind expected to do?
YD1	The wind is expected to back.
YD2	The wind is expected to veer.
YD3	The wind is expected to increase.
YD4	The wind is expected to become squally.
YD5	The wind is expected to remain steady in force.
YD6	The wind is expected to moderate.
YG	You appear not to be complying with the traffic regulation scheme.

COMMUNICATIONS

ACKNOWLEDGE—ANSWER.

YH	I have received the following from... (name or identity signal of vessel or station).
YI	I have received the safety signal sent by... (name or identity signal).
YJ	Have you received the safety signal sent by... (name or identity signal)?
YK	I am unable to answer your question.
	Received, or I have received your last signal. (Pro. Sig.) *R*

CALLING.

YL	I will call you again at... hours (on... kHz (or MHz).
YM	Who is calling me?

CANCEL.

YN	Cancel my last signal/message.
	My last signal was incorrect. I will repeat it correctly.. *ZP*

COMMUNICATE.

	I wish to communicate with you by ... (Complements Table 1). (With one numeral). .... *K*
	I wish to communicate with you. .... *K*
YO	I am going to communicate by ... (Complements Table I).
YP	I wish to communicate with vessel or coast station (identity signal) by ... (Complements Table I).
YQ	I wish to communicate by ... (Complements Table I) with vessel bearing ... from me.
YR	Can you communicate by ... (Complements Table I)?
YS	I am unable to communicate by ... (Complements Table I).
YU[1]	I am going to communicate with your station by means of the International Code of Signals.
YV[1]	The groups which follow are from the International Code of Signals.
YV1	The groups which follow are from the Local Code.
YW	I wish to communicate by radiotelegraphy on frequency indicated.
YX	I wish to communicate by radiotelephony on frequency indicated.
YY	I wish to communicate by VHF radiotelephony on channels indicated.
YZ	The words which follow are in plain language.
ZA	I wish to communicate with you in ... (language indicated by the following Complement):

      0 Dutch   **3** German   **6** Japanese
      1 English   **4** Greek   **7** Norwegian
      2 French   **5** Italian   **8** Russian
                                            9 Spanish

ZB	I can communicate with you in language indicated (Complements as above).
ZC	Can you communicate with me in language indicated (complements as above)?
ZD	Please communicate the following to all shipping in the vicinity.
ZE	You should come within visual signal distance.
	You should keep within visual signal distance from me (or vessel indicated). *PR*2

---

[1] The abbreviation INTERCO may also be used to mean International Code group(s) follow(s).

	I have established communications with aircraft in distress on 2182 kHz.....*BC*
	Can you communicate with aircraft.....*BC*1
	I have established communications with the aircraft in distress on ... MHz.....*BE*

EXERCISE.

ZF	I wish to communicate with you by...(Complements Table I).
ZG	It is not convenient to excercise signals.
ZH	Exercise has been completed.

RECEPTION—TRANSMISSION.

ZI	I can receive but not transmit by...(Complements Table I).
ZJ	I can transmit but not receive by ...(Complements Table I).
ZK	I cannot distinguish your signal. Please repeat it by...(Complements Table I).
ZL	Your signal has been received but not understood. I cannot read your...(Complements Table I). ....*YT*
ZM	You should send (or speak) more slowly.
ZM1	Shall I send (or speak) more slowly?
ZN	You should send each word or group twice.
ZO	You should stop sending.
ZO1	Shall I stop sending?

REPEAT.

ZP	My last signal was incorrect. I will repeat it correctly.
ZQ	Your signal appears incorrectly coded. You should check and repeat the whole.
ZR	Repeat the signal now being made to me by vessel (or coast station)...(name or identity signal).

INTERNATIONAL SANITARY REGULATIONS.
PRATIQUE MESSAGES.

ZS	My vessel is 'healthy' and I request free pratique.......*Q*
	[1]I require health clearance.........*QQ*

---

[1] By Night—A Red light over White light may be shown (2m apart).

ZT	My Maritime Declaration of Health has negative answers to the six health questions.
ZU	My Maritime Declaration of Health has a positive answer to question(s) ... (indicated by number(s)).
ZV	I believe I have been in an infected area during the last thirty days.
ZW	I require Port Medical Officer.
ZW1	Port Medical Officer will be available at (time indicated).
ZX	You should make the appropriate pratique signal.
ZY	You have pratique.
ZZ	You should proceed to anchorage for health clearance at (place indicated).
ZZ1	Where is the anchorage for health clearance?

I have a doctor on board. .... *AL*
Have you a doctor? .... *AM*

TABLES OF COMPLEMENTS

Table I.

1. Semaphore.
2. Morse signalling by handflags or arms.
3. Loud hailer (megaphone).
4. Morse signalling lamp.
5. Sound signals.
6. International Code flags.
7. Radiotelegraphy 500 kHz.
8. Radiotelephony 2182 kHz.
9. VHF Radiotelephony-channel 16.

Table II.

0. Water.
1. Provisions.
2. Fuel.
3. Pumping equipment.
4. Fire-fighting appliances.
5. Medical assistance.
6. Towing.
7. Survival craft.
8. Vessel to stand by.
9. Icebreaker.

Table III.
   0. Direction unknown (or calm).
   1. North-east.
   2. East.
   3. South-east.
   4. South.
   5. South-west.
   6. West.
   7. North-west.
   8. North.
   9. All directions (or confused or variable).

CHAPTER XIII

MEDICAL SECTION
General Instructions

1. Medical advice should be sought and given in plain language whenever possible but, if language dificulties are encountered, this code should be used.
2. Even when plain language is used, the text of the code and the instructions should be followed as far as possible.
3. Reference is made to the procedure signals C, N, NO or RQ which, when used after the main signal, change its meaning into affirmative, negative and interrogative respectively.

Examples—MFB N = 'Swelling is not discharging.'
MFI RQ = 'Has bleeding stopped.'

INSTRUCTIONS TO MASTERS
Standard Method of Case Description

1. The Master should make a careful examination of the patient and should try to collect, as far as possible, information covering the following subjects:—

(a) Description of patient.
(b) Previous health.
(c) Location of symptoms, diseases or injuries.
(d) General symptoms.
(e) Diagnosis.

Examples— MQE 32 = 'My probable diagnosis is duodenal ulcer.'

Prescribing MRR 22 = 'Apply hot water bottle to lower abdomen.'

Prescribing a medicant MTD 18 = 'You should give Penicillin injection.'

Examples:
Method of Administration and Dose.
MT12 = 'You should give by mouth two tablets/capsules.'

Frequency of Dose.
MTP4 = 'You should repeat after 4 hours.'
Diet Advice.
MUB = 'Give water very freely.'

PART I

REQUEST FOR MEDICAL ASSISTANCE

Request—General Information.

MAA	I request urgent medical Advice.
MAB	I request you make rendezvous in position indicated.
MAC	I request you to arrange hospital admission.
MAD	I am ... (indicate number) hours from nearest port.

Description of Patient.

MAJ	I have a male aged ... (number) years.
MAK	I have a female aged ... (number) years.
MAL	I have a female ... (number) months pregnant.
MAM	Patient has been ill for ... (number) days.
MAN	Patient has been ill for ... (number) hours.
MAO	General condition of the patient is good.
MAP	General condition of the patient is serious.
MAQ	General condition of the patient is unchanged.
MAR	General condition of the patient has worsened.
MAS	Patient has been given ... (Table MIII) with effect.
MAT	Patient has been given ... (Table MIII) without effect.
MBA	Patient has suffered from ... (Table MII).
MBB	Patient has had previous operation ... (Table MII).

Location of Symptoms, Diseases or Injuries

MBE	The whole body is affected.
MBF	The part of the body affected is ... (Table MI).
MBG	The part of the body affected is right ... (Table MI).
MBH	The part of the body affected is left ... (Table MI).

General Symptoms.

MBP	Onset was sudden.
MBQ	Onset was gradual.

Temperature.

MBR	Temperature taken in the mouth is . . . (number).
MBT	Temperature taken in the morning is . . . (number).
MBV	Temperature is rising.
MBW	Temperature is falling.

Pulse.

MBX	The pulse rate per minute is . . . (number).
MBY	The pulse rate is irregular.
MBZ	The pulse rate is rising.
MCA	The pulse rate is falling.
MCB	The pulse rate is weak.

Breathing.

MCE	The rate of breathing per minute is . . . (number) (in and out being counted as one breath).
MCF	The breathing is weak.
MCH	The breathing is regular.
MCI	The breathing is irregular.

Sweating.

MCL	Patient is sweating.
MCM	Patient has fits of shivering (chills).
MCO	Patient's skin is hot and dry.
MCP	Patient is cold and clammy.

Mental State and Consciousness.

MCR	Patient is conscious.
MCT	Patient is semi-conscious but can be roused.
MCU	Patient is unconscious.
MCV	Patient found unconscious.
MCW	Patient appears to be in a state of shock.
MCX	Patient is delirious.
MCZ	Patient is paralyzed . . . (Table MI).
MDD	Patient is unable to sleep.

Pain.

MDF	Patient is in pain ... (Table MI).
MDG	Pain is dull ache.
MDJ	Pain is slight.
MDL	Pain is severe.
MDN	Pain is continuous.
MDO	Pain is increased by hand pressure.
MDP	Pain radiates to ... (Table MI).
MDQ	Pain is increased on breathing.
MDR	Pain is increased by action of bowels.
MDX	Pain is relieved by heat.
MDY	Pain has ceased.

Cough.

MED	Cough is present.
MEF	Cough is absent.

Bowels.

MEG	Bowels are regular.
MEJ	Patient is constipated and bowels last opened ... (indicate number of days).

Vomitting.

MEM	Vomiting is present.
MEN	Vomiting is absent.

Urine.

MEP	Urinary functions normal.
MEQ	Urinary functions abnormal.

Bleeding

MER	Bleeding is present ... (Table MI).
MET	Bleeding is absent.

Rash.

MEU	A rash is present ... (Table MI).
MEV	A rash is absent.

Swelling.

MEW	Patient has a swelling...(Table MI).
MEX	Swelling is hard.
MEY	Swelling is soft.
MFB	Swelling is discharging.
MFC	Patient has an abscess...(Table MI).
MFD	Patient has a carbuncle...(Table MI).

PARTICULAR SYMPTOMS

Accidents, Injuries, Fractures, Suicide and Poisons

MFE	Bleeding is severe.
MFF	Bleeding is slight.
MFG	Bleeding has been stopped by pad(s) and bandaging.
MFH	Bleeding has been stopped by tourniquet.
MFI	Bleeding has stopped.
MFJ	Bleeding cannot be stopped.
MFK	Patient has a superficial wound...(Table MI).
MFL	Patient has a deep wound...(Table MI).
MFM	Patient has a penetrating wound...(Table MI).
MFQ	Patient has contusion (bruising)...(Table MI)
MFR	Wound is due to blow.
MFS	Wound is due to crushing.
MFT	Wound is due to explosion.
MFU	Wound is due to fall.
MFV	Wound is due to gun-shot.
MFW	Patient has a foreign body in wound.
MFX	Patient is suffering from concussion.
MFY	Patient cannot move the arm...(Table MI).
MFZ	Patient cannot move the leg...(Table MI).
MGA	Patient has dislocation...(Table MI).
MGB	Patient has simple fracture...(Table MI).
MGC	Patient has compound fracture...(Table MI).
MGE	Patient has attempted suicide.
MGH	Patient has severe burn...(Table MI).
MGI	Patient is suffering from non-corrosive poisoning (no staining and burning of mouth and lips).

MGJ	Patient has swallowed corrosive (staining and burning of mouth and lips).
MGM	Emetic has been given with good results
MGO	No emetic has been given.
MGR	Patient is suffering from animal bite...(Table MI).
MGS	Patient is suffering from snake bite...(Table MI).
MGT	Patient is suffering from gangrene...(Table MI).

Diseases of Nose and Throat

MGU	Patient has nasal discharge.
MHA	Lips are swollen.
MHF	Tongue is swollen.
MHL	Throat is sore and red.
MHO	Throat hurts and is swollen on one side.
MHQ	Swallowing is painful.
MHR	Patient cannot swallow. Patient has swallowed a foreign body..............*MGL*
MHV	Patient has severe toothache.

Diseases of Respiratory System

MHY	Patient has pain in chest on breathing...(Table MI). Breathing is wheezing..........................*MCG*
MHZ	Breathing is deep.
MIB	Patient has asthmatical attack.
MIC	Patient has severe cough.
MIF	Patient is coughing up blood.
MIL	Patient has bloodstained sputum.

Diseases of the Digestive System

MIN	Patient has tarry stool.
MIQ	Patient is passing blood with stools.
MIU	Patient has cramp pains and vomitting.
MIV	Vomiting has stopped.
MIX	Vomit is streaked with blood.
MJI	Abdominal wall is hard and rigid.
MJJ	Abdominal wall is tender...(Table MI).

MJK	Hernia is present.
MJO	Patient has bleeding haemorrhoids.

Diseases of the Geniti-Urinary System

MJS	Patient has pain on passing water.
MJV	Patient is unable to hold urine (incontinent).
MJY	Amount of urine passed in 24 hours ... (indicate number in decilitres; 1 decilitre equals one-sixth of a pint).
MKA	Urine contains albumen.
MKB	Urine contains sugar.
MKC	Urine contains blood.
MKJ	Shall I pass a catheter?
MKK	I have passed a catheter.

Diseases of the Nervous System and Mental Diseases

MKP	Patient has headache ... (Table MI).
MKR	Headache is very severe.
MKU	Patient is unable to speak properly.
MKW	Pupils are equal in size.
MKX	Pupils are unequal in size.
MLD	Patient is uncontrollable.
	Patient has attempted suicide.....................*MGE*
MLF	Patient has delirium tremens.

Diseases of the Heart and Circulatory System

	Patient is in pain ... (Table MI).................*MDF*
MLH	Pain has been present for ... (indicate number of minutes).
MLI	Pain in chest in constricting character.
MLJ	Pain is behind the breastbone.
MLL	Breathing is difficult when lying down.

Infectious and Parasitic Diseases

MLR	Rash has been present for ... (indicate number of hours).
MLS	Rash first appeared on ... (Table MI).
MLT	Rash is spreading to ... (Table MI).
MMB	Rash is weeping (oozing).

MMC	Rash looks like weals.
MME	Skin is yellow.
MMJ	Patient has been isolated.
	Patient has never been successfully vaccinated against smallpox.................................*MUT*

Veneral Disease (See also Diseases of Genito-Urinary System)

MMQ	Patient has previous history of gonorrhoea.
	Patient has buboes... (Table MI) .... *MMF*
MMT	Patient has swollen glands in the groin.

Diseases of the Ear

MMW	Patient has boil in ear(s).
MMZ	Patient has discharge of pus from ear(s).
MNB	Patient has foreign body in ear.

Diseases of the Eye.

MNG	Patient has inflammation of eye(s).
MNH	Patient has discharge from eye(s).
MNI	Patient has foreign body embedded in the pupil area of the eye.
MNN	Patient has sudden blindness in one eye.
MNO	Patient has sudden blindness in both eyes.

Diseases of the Skin.

See Infectious and Parasitic Diseases.

Diseases of Muscles and Joints.

MNT	Patient has pain in muscles of... (Table MI).
MNU	Patient has pain in joint(s)... (Table MI).
MNV	Patient has redness and swelling of joint(s) (Table MI).
MNW	There is history of recent injury.
MNX	There is no history of injury.

Miscellaneous Illnesses.

MOA	Patient is suffering from heat exhaustion.
MOB	Patient is suffering from heat-stroke.
MOC	Patient is suffering from seasickness.

MOD	Patient is suffering from exposure in lifeboat—Indicate length of exposure (number) hours.
MOE	Patient is suffering from frostbite... (Table MI).
MOF	Patient has been exposed to radioactive hazard.

Childbirth.

MOK	I have a patient in childbirth aged... (number) years.
MOL	Patient states she has had... (number) children.
MOM	Patient states child is due in... (number) weeks.
MON	Pains began... (number) hours ago.
MOQ	Pains are occurring every... (number) minutes.
MOR	The bag of membrane broke... (number) hours ago.
MOS	There is severe bleeding from womb.
MOX	The child has been born.
MOY	The child will not breathe.
MOZ	The placenta has been passed.
MPA	The placenta has not been passed.

Progress Report.

MPE	I am carrying out prescribed instructions.
MPF	Patient is improving.
MPG	Patient is not improving.
MPJ	Patient is restless.
MPK	Patient is calm.
MPL	Symptoms have cleared.
MPM	Symptoms have not cleared.
MPN	Symptoms have increased.
MNO	Symptoms have decreased.
MPP	Treatment has been effective.
MPQ	Treatment has been ineffective.
MPR	Patient has died.

PART II—MEDICAL ADVICE

Request for Additional Information.

MQB	I cannot understand your signal; please use standard method of case description.
MQC	Please answer the following question(s).

Diagnosis.

MQE	My probable diagnosis is ... (Table MII).
MQF	My alternative diagnosis is ... (Table MII).
MQG	My probable diagnosis is infection or inflammation ... (Table MI).
MQH	My probable diagnosis is perforation of ... (Table MI).
MQI	My probable diagnosis is tumour of ... (Table MI).
MQJ	My probable diagnosis is obstruction of ... (Table MI).
MQK	My probable diagnosis is haemorrhage of ... (Table MI).
MQL	My probable diagnosis is foreign body in ... (Table MI).
MQM	My probable diagnosis is fracture of ... (Table MI).
MQN	My probable diagnosis is dislocation of ... (Table MI).
MQO	My probable diagnosis is sprain of ... (Table MI).
MQP	I cannot make a diagnosis.
MQT	Your diagnosis is probably right.
MQU	I am not sure about your diagnosis.

Special Treatment.

MRI	You should refer to your *International Ship's Medical Guide* if available or its equivalent.
MRQ	Apply hot compress and renew every ... (number) hours.
MRR	Apply hot water bottle to ... (Table MI).
MRT	Insert antiseptic eye drops ... (number) times daily.
MRW	Give frequent gargles one teaspoonful of salt in a tumblerful of water.
MRX	Give enema.
MRY	Do not give enema or laxative.
MSD	Apply cotton wool to armpit and bandage arm to side.
MSF	Apply a sling and/or rest the part.
MSL	Reduce temperature of patient as indicated in general nursing chapter.
MSO	Dress wound with sterile gauze, cotton wool and apply well-padded splint.
MSP	Apply burn and wound dressing and bandage lightly.
MSU	Stop bleeding by applying more cotton wool, firm bandaging and elevation of the limb.
MSV	Stop bleeding by manual pressure.

MSW	Apply tourniquet for not more than fifteen minutes.
MSY	You should pass a stomach tube.
MSZ	Do not try to empty stomach by any method.

Treatment by Medicaments.

Prescribing.

MTD	You should give ... (Table MIII).
MTE	You must not give ... (Table MIII).

Method of Administration and Dose.

MTF	You should give one tablespoonful (15 ml. or ½ oz.).
MTG	You should give dessertspoonful (7·5 ml. or ¼ oz.).
MTH	You should give one teaspoonful (4 ml. or 1 drachm.).
MTI	You should give by mouth ... (number) tablets/capsules.
MTJ	You should give a tumblerful of water with each dose.
MTK	You should give by intramuscular injection ... (number) milligrammes.
MTL	You should give by subcutaneous injection ... (number) milligrammes.
MTM	You should give by intramuscular injection ... (number) ampoule(s).
MTN	You should give by subcutaneous injection ... (number) ampoule(s).

Frequency of Dose.

MTO	You should give once only.
MTP	You should repeat after ... (number) hours.
MTQ	You should repeat every ... (number) hours.
MTR	You should continue for ... (number) hours.

Frequency of External Application.

MTT	You should apply once only.
MTU	You should apply every ... (number) hours.
MTV	You should cease to apply.
MTW	You should apply for ... (number) minutes.

Diet.

MUA	Give nothing by mouth.
MUB	Give water very freely.

MUC	Give water only in small quantities.
MUF	Give fluid diet, milk, fruit juices, tea, mineral water.
MUG	Give light diet such as vegetable soup, steamed fish, stewed fruit, milk puddings or equivalent.
MUH	Give normal diet as tolerated.

Childbirth.

MUI	Has she had previous children?
MUJ	How many months pregnant is she?
MUK	When did labour pains start?
MUM	Encourage her to strain down during pains.
MUN	What is the frequency of pains (indicate in minutes)?
MUP	You should apply tight wide binder around lower part of abdomen and hips.
MUQ	You should apply artificial respiration gently by mouth technique on infant.

Vaccination Against Smallpox.

MUR	Has the patient been successfully vaccinated?
MUS	Has the patient been vaccinated during the past three years?
MUT	Patient has never been successfully vaccinated against smallpox.
MUU	Patient was last vaccinated ... (indicate date).

General Instructions.

MVA	I consider the case is serious and urgent.
MVB	I do not consider the case serious or urgent.
MVG	Keep patient warm.
MVH	Keep patient cool.
MVI	You should continue your local treatment.
MVK	You should continue giving ... (Table MIII).
MVL	You should suspend your local treatment.
MVM	You should suspend your special treatment.
MVN	You should cease giving ... (Table MIII).
MVP	You should land your patient at the earliest opportunity.
MVR	I will arrange for hospital admission.
MVT	No treatment advised.
MVU	Refer back to me in ... (number) hours or before if patient worsens.

TABLE OF COMPLEMENTS

TABLE MI

Regions of the Body

Side of body or limb affected should be clearly indicated—right, left.

FRONT

01. Frontal region of head
02. Side of head
03. Top of head
04. Face
05. Jaw
06. Neck Front
07. Shoulder
08. Clavicle
09. Chest
10. Chest, mid
11. Heart
12. Armpit
13. Arm, upper
14. Forearm
15. Wrist
16. Palm of hand
17. Fingers
18. Thumb
19. Central upper abdomen
20. Central lower abdomen
21. Upper abdomen
22. Lower abdomen
23. Lateral abdomen
24. Groin
25. Scrotum
26. Testicles
27. Penis
28. Upper thigh
29. Middle thigh
30. Lower thigh
31. Knee
32. Patella
33. Front of leg
34. Ankle
35. Foot
36. Toes

BACK

37. Back of head
38. Back of neck
39. Back of shoulder
40. Scapula region
41. Elbow
42. Back upper arm
43. Back lower arm
44. Back of hand
45. Lower chest region
46. Spinal column, upper
47. Spinal column, middle
48. Spinal column, lower
49. Lumber (kidney) region
50. Sacral region
51. Buttock
52. Anus
53. Back of thigh
54. Back of knee
55. Calf
56. Heel

Other Organs of the Body

57. Artery
58. Bladder
59. Brain
60. Breast
61. Ear(s)
62. Eye(s)
63. Eyelids
64. Gall bladder
65. Gullet (oesophagus)
66. Gums
67. Intestine
68. Kidney
69. Lip, lower
70. Lip, upper
71. Liver
72. Lungs
73. Mouth
74. Nose
75. Pancreas
76. Prostate
77. Rib(s)
78. Spleen
79. Stomach
80. Throat
81. Tongue
82. Tonsils
83. Tooth, Teeth
84. Urethra
85. Uterus, womb
86. Vein
87. Voice box (larynx)
88. Whole abdomen
89. Whole arm
90. Whole back
91. Whole chest
92. Whole leg

TABLE MII
List of common Diseases

1. Abscess
2. Alcoholism
3. Allergic reaction
4. Amoebic dysentery
5. Angina pectoris
6. Anthrax
7. Apoplexy (stroke)
8. Appendicitis
9. Asthma
10. Bacillary dysentery
11. Boils
12. Bronchitis (acute)
13. Bronchitis (chronic)
14. Brucellosis
15. Carbuncle
16. Cellutitis
17. Chancroid
18. Chickenpox
19. Cholera
20. Cirrhosis of the liver
21. Concussion
22. Compression of brain
23. Congestive heart failure
24. Constipation
25. Coronary thrombosis
26. Cystitis (bladder inflammation)
27. Dengue
28. Diabetes
29. Diabetic coma
30. Diptheria
31. Drug re-action
32. Duodenal ulcer
33. Eczema
34. Erysipelas
35. Fits
36. Gangrene
37. Gastric Ulcer
38. Gastro-enteritis
39. Gonorrhoea

40. Gout
41. Heat cramps
42. Heat exhaustion
43. Heat-stroke
44. Hepatitis
45. Hernia
46. Hernia (irreducible)
47. Hernia (strangulated)
48. Immersion foot
49. Impetigo
50. Insulin overdose
51. Indigestion
52. Influenza
53. Intestinal obstruction
54. Kidney stone (renal colic)
55. Laryngitis
56. Malaria
57. Measles
58. Meningitis
59. Mental illness
60. Migraine
61. Mumps
62. Orchitis
63. Peritonitis
64. Phlebitis
65. Piles
66. Plague
67. Pleurisy
68. Pneumonia
69. Poisoning (corrosive)
70. Poisoning (non-corrosive)
71. Poisoning (barbiturates)
72. Poisoning (methyl alcohol)
73. Poisoning (gases)
74. Poliomyelitis
75. Prolapsed inter-vertebral disc (slipped disc)
76. Pulmonary tuberculosis
77. Quinsy
78. Rheumatism
79. Rheumatic fever
80. Scarlet fever
81. Sciatica
82. Shingles (herpes zoster)
83. Sinusitis
84. Shock
85. Smallpox
86. Syphylis
87. Tetanus
88. Tonsillitis
89. Typhoid
90. Typhus
91. Urethritis
92. Urticaria (nettle rash)
93. Whooping cough
94. Yellow fever

TABLE MIII

LIST OF MEDICANTS

A. For External Use

01. Ear drops
02. Antiseptic eye drops
03. Anaesthetic eye drops
04. Salicylate liniment
05. Calamine lotion
06. Antiseptic lotion
07. Nasal drops
08. Soft paraffin
09. Burn/wound dressing
10. Antibiotic ointment
11. Pile ointment
12. Local Anaesthetic ointmen

B. For Internal Use

13. Antihistamine tablets
14. Adrenaline (1 mg. in 'Ampins')

Antibiotics
15. Tetracycline capsules (250 mg. per capsule)
16. Penicillin tablets (125 mg. per tablet)
17. Sulfonamide tablets (500 mg. per tablet)
18. Penicillin injection (600,000 units per ampoule)
19. Streptomycin injection (1,000 mg. per ampoule)
20. Tetracycline injection (100 mg. per ampoule)

Asthma
21. Asthma relief tablets (300 mg. per tablet)
22. Ephedrine tablets (30 mg. per tablet)
23. Inhalation mixture

Cough
24. Codeine tablets (15 mg. per tablet)
25. Cough linctus
26. Diarrhoea mixture

Heart
27. Heart tablets (0·5 mg. per tablet)

NOTE—*For congestive heart failure the following preparation are available on board ship but they should be used only on medical advice transmitted in plain language and not by code:* Chlorothiazide or equivalent (500 mg. per tablet); Digozin tablets or equivalent (0·25 mg. per tablet)

Indigestion
28. Stomach tablets

Laxatives
29. Vegetable laxative tablets.
30. Liquid laxative—'Milk of Magnesia'

Malaria
31. Malaria tablets (200 mg. per tablet)

Pain
32. Asprin tablets (300 mg. per tablet)
33. Morphine injection (15 mg. per ampoule)

Sedation
34. Sedative tablets (100 mg. per tablet)
35. Phenobarbitone tablets (30 mg. per tablet)
36. Tranquillizer tablets (Largactil) (50 mg. per tablet)

Caution: This tablet No. 36 to be used only on medical advice by radio

Salt Depletion or Heat Cramps
37. Salt tablets (500 mg. per tablet).

Seasickness
38. Seasickness tablets (0·3 mg. per tablet)

SHIPS IN DISTRESS

Statutory distress signals

Annex IV of the International Regulations for Preventing Collisions at Sea 1972, lists the signals to be used or exhibited either together or separately to indicate distress and need of assistance. These are:—

(1) *(a)* a gun or other explosive signal fired at intervals of about a minute;

(b) a continuous sounding with any fog-signalling apparatus;

(c) rockets or shells, throwing red stars fired one at a time at short intervals;

(d) a signal made by radiotelegraphy or by any other signalling method consisting of the group ▄ ▄ ▄ ▬ ▬ ▬ ▄ ▄ ▄ (SOS) in the Morse Code;

(e) a signal sent by radiotelephony consisting of the spoken word "Mayday";

(f) the International Code Signal of distress indicated by N.C.;

(g) a signal consisting of a square flag having above or below it a ball or anything resembling a ball;

(h) flames on the vessel (as from a burning tar barrel, oil barrel, etc);

(i) a rocket parachute flare or a hand flare showing a red light.

(j) a smoke signal giving off orange-coloured smoke;

(k) slowly and repeatedly raising and lowering arms outstretched to each side;

(l) the radiotelegraph alarm signal;

(m) the radiotelephone alarm signal;

(n) signals transmitted by emergency position-indicating radio beacons.

(2) The use or exhibition of any of the foregoing signals except for the purpose of indicating distress and need of assistance and the use of other signals which may be confused with any of the above signals is prohibited.

(3) Attention is drawn to the relevant sections of the International Code of Signals, the Merchant Ship Search and Rescue Manual and the following signals;

(a) a piece of orange-coloured canvas with either a black square and circle or other appropriate symbol (for identification from the air);

(b) a dye marker.

See page 195 for Aircraft in Distress.

TABLE SHOWING INTERNATIONAL ALLOCATION OF INITIAL LETTERS OF SIGNAL LETTERS, CALL SIGNS AND AIRCRAFT MARKINGS

United States of America	AAA—ALZ
(Not allocated)	AMA—AOZ
Pakistan	APA—ASZ
India	ATA—AWZ
Commonwealth of Australia	AXA—AXZ
Argentina Republic	AYA—AZZ
China	BAA—BZZ
Chile	CAA—CEZ
Canada	CFA—CKZ
Cuba	CLA—CMZ
Morocco	CNA—CNZ
Cuba	COA—COZ
Bolivia	CPA—CPZ
Portuguese Colonies	CQA—CRZ
Portugal	CSA—CUZ
Uruguay	CVA—CXZ
Canada	CYA—CZZ
Germany	DAA—DMZ
Belgian Congo	DNA—DQZ
Bielorussian Soviet Socialist Republic	DRA—DTZ
Republic of the Philippines	DUA—DZZ
Spain	EAA—EHZ
Ireland	EIA—EJZ
Union of Soviet Socialist Republics	EKA—EKZ
Republic of Liberia	ELA—ELZ
Union of Soviet Socialist Republics	EMA—EOZ
Iran	EPA—EQZ
Union of Soviet Socialist Republics	ERA—ERZ
Estonia	ESA—ESZ
Ethiopia	ETA—ETZ
Union of Soviet Socialist Republics	EUA—EZZ
France and Colonies and Protectorates	FAA—FZZ
Ghana	G
Great Britain	GAA—GZZ
Hungary	HAA—HAZ
Switzerland	HBA—HBZ
Ecuador	HCA—HDZ
Switzerland	HEA—HEZ
Poland	HFA—HFZ
Hungary	HGA—HGZ

Republic of Haiti	HHA—HHZ
Dominican Republic	HIA—HIZ
Republic of Colombia	HJA—HKZ
Korea	HLA—HMZ
Iraq	HNA—HNZ
Republic of Panama	HOA—HPZ
Republic of Honduras	HQA—HRZ
Siam	HSA—HSZ
Nicaragua	HTA—HTZ
Republic of El Salvador	HUA—HUZ
Vatican City State	HVA—HVZ
France and Colonies and Protectorates	HWA—HYZ
Kingdom of Saudi Arabia	HZA—HZZ
Italy and Colonies	IAA—IZZ
Japan	JAA—JSZ
Mongolian Peoples' Republic	JTA—JVZ
Norway	JWA—JXZ
(Not allocated)	JYA—JZZ
United States of America	KAA—KZZ
Norway	LAA—LNZ
Argentina Republic	LOA—LWZ
Luxemburg	LXA—LXZ
Lithuania	LYA—LYZ
Bulgaria	LZA—LZZ
Great Britain	MAA—MZZ
United States of America	NAA—NZZ
Peru	OAA—OCZ
Republic of Lebanon	ODA—ODZ
Austria	OEA—OEZ
Finland	OFA—OJZ
Czechoslovakia	OKA—OMZ
Belgium and Colonies	ONA—OTZ
Denmark	OUA—OZZ
Netherlands	PAA—PIZ
Curacao	PJA—PJZ
Netherland Indies	PKA—POZ
Brazil	PPA—PYZ
Surinam	PZA—PZZ
(Service abbreviations)	QAA—QZZ
Union of Soviet Socialist Republics	RAA—RZZ
Sweden	SAA—SMZ
Poland	SNA—SRZ

Egypt	SSA—SUZ
Greece	SVA—SZZ
Turkey	TAA—TCZ
Guatemala	TDA—TDZ
Costa Rica	TEA—TEZ
Iceland	TFA—TFZ
Guatemala	TGA—TGZ
France and Colonies and Protectorates	THA—THZ
Costa Rica	TIA—TIZ
France and Colonies and Protectorates	TJA—TZZ
Union of Soviet Socialist Republics	UAA—UQZ
Ukrainian Soviet Socialist Republic	URA—UTZ
Union of Soviet Socialist Republics	UUA—UZZ
Canada	VAA—VGZ
Commonwealth of Australia	VHA—VNZ
Newfoundland	VOA—VOZ
British Colonies and Protectorates	VPA—VSZ
India	VTA—VWZ
Canada	VXA—VYZ
Commonwealth of Australia	VZA—VZZ
United States of America	WAA—WZZ
Mexico	XAA—XIZ
Canada	XJA—XOZ
Denmark	XPA—XPZ
Chile	XQA—XRZ
China	XSA—XSZ
France and Colonies and Protectorates	XTA—XWZ
Portuguese Colonies	XXA—XXZ
Burma	XYA—XZZ
Afghanistan	YAA—YAZ
Netherlands Indies	YBA—YHZ
Iraq	YIA—YIZ
New Hebrides	YJA—YJZ
Syria	YKA—YKZ
Latvia	YLA—YLZ
Turkey	YMA—YMZ
Nicarugua	YNA—YNZ
Rumania	YOA—YRZ
Republic of El Salvador	YSA—YSZ
Yugoslavia	YTA—YUZ
Venezuela	YYA—YYZ
Yugoslavia	YZA—YZZ
Albania	ZAA—ZAZ

British Colonies and Protectorates	ZBA—ZJZ
New Zealand	ZKA—ZMZ
British Colonies and Protectorates	ZNA—ZOZ
Paraguay	ZPA—ZPZ
British Colonies and Protectorates	ZQA—ZQZ
Union of South Africa	ZRA—ZUZ
Brazil	ZVA—ZZZ
Great Britain	2AA—2ZZ
Principality of Monaco	3AA—3AZ
Canada	3BA—3FZ
Chile	3GA—3GZ
China	3HA—3UZ
France and Colonies and Protectorates	3VA—3VZ
(Not allocated)	3WA—3XZ
Norway	3YA—3YZ
Poland	3ZA—3ZZ
Mexico	4AA—4CZ
Republic of the Philippines	4DA—4IZ
Union of Soviet Socialist Republics	4JA—4LZ
Venezuela	4MA—4MZ
Yugoslavia	4NA—4OZ
British Colonies and Protectorates	4PA—4SZ
Peru	4TA—4TZ
United Nations	4UA—4UZ
Republic of Haiti	4VA—4VZ
Yemen	4WA—4WZ
Kenya	5Z
Malaya	9M

NOTE—The following call signs have been allotted by the Administrative Council of the International Telecommunications Union until the Radio Regulations as revised by the next Administrative Radio Conference come into force:—

State of Israel	4XA—4XZ
French Protectorate of Morocco	5CA—5CZ
San Marino	9AA—9AZ
Nepal	9NA—9NZ

SIGNAL LETTERS

The following list of signal letters is inserted for use with the exercises given in this book, and is not to be relied upon for signalling at sea.

Name of Vessel	Signal Letters	Country
Baron Forbes	G B M R	G.
British Advocate	G J L M	G.
British Captain	G J R N	G.
British Diplomat	G F R Y	G.
City of Agra	G R B C	G.
City of Calcutta	G C P K	G.
City of Exeter	G Q Z W	G.
City of Kimberley	G L C M	G.
Daylight	W D E J	U.S.A.
Earl Haig	G Q M J	G.
Elmpark	G D M F	G.
Euripides	G M L P	G.
Harvard	W T B Y	U.S.A.
Hilary	G Q V M	G.
Holiday	W F E W	U.S.A
Loch Leven	G T B Y	G.
Mateba	O T B A	Belg.
Northern Light	K G E G	U.S.A.
Opawa	G J M C	G.
Southern Star	W G E B	U.S.A.

CHAPTER XIV

FIRING PRACTICE AND EXERCISE AREAS

Firing and bombing practices, and defence exercises, take place in a number of areas in Home Waters and off the coasts of British Commonwealth and Colonial Territories as well as in foreign waters.

In future, and in view of the responsibility of range authorities to avoid accidents, limits of practice areas (with the exception of Submarine Exercise Areas) will not as a rule be shown on navigational charts and descriptions will not appear in the Sailing Directions. They are, however, shown in Home Waters on a series of six small scale charts called the PEXA series.

Such range beacons, lights and marking buoys as may be of assistance to the mariner, or targets which might be a danger to navigation, will, however, be indicated on navigational charts and, when appropriate, mentioned in Sailing Directions. Lights will be mentioned in the Admiralty List of Lights.

The principal types of practices carried out are:—
(a) Bombing practice from aircraft.
  Warning signals are usually shown.
(b) Air to air, and air to sea or ground firing.

The former is carried out by aircraft at a large white or red sleeve, a winged target or flag towed by another aircraft moving on a steady course. The latter is carried out from aircraft at towed or stationary targets on sea or land, the firing taking place to seaward in the case of those on land.

As a general rule, warning signals are shown when the targets are stationary, but not when towed targets are used.

All marine craft operating as range safety craft, target towers or control launches for wireless controlled target will display, for identification purposes, while in or in the vicinity of the danger area, the following markings:—
  (1) A large red flag at the masthead;
  (2) A painted canvas strip, 1·83 metres by 0·92 metres with red and white chequers in 39 centimetre squares, on the fore deck or cabin roof.

(c) Anti-aircraft firing.
   This may be from A.A. guns or machine guns at a target towed by aircraft as in *(b)* above, a pilotless target aircraft, or at balloons or kites. Practice may take place from shore batteries or ships.
   Warning signals as a rule are shown from shore batteries. Ships fly a red flag

(d) Firing from shore batteries or ships at sea at fixed or floating targets.
   Warning signals are usually shown as in *(c)*.

(e) At remote-controlled craft.
   These craft are 19·20 metres in length and carry "not under command" shapes and lights. Exercises consisting of surface firing by ships, practice bombing, air to sea firing and rocket firing will be carried out against these craft or targets towed by them.
   A control craft will keep visual and radar watch up to approximately 8 miles and there will be cover from the air over a much greatr range to ensure that other shipping will not be endangered.

(f) Rocket and Guided Weapons firing.
   These may take the form of *(b)*, *(c)* or *(d)* above. All such firings are conducted under Clear (Air and Sea) Range procedure. Devices are generally incorporated whereby the missiles may be destroyed should their flights be erratic.
   Warning signals are usually shown as in *(c)* above.

Warning signals, when given, usually consist of red flags by day and *red fixed or red flashing* lights at night. The absence of any such signal cannot, however, be accepted as evidence that a practice area does not exist. Warning signals are shown from shortly before practice commences until it ceases.

Ships and aircraft carrying out exercises may illuminate with bright coloured flares. To avoid confusion with international distress signals, red or orange flares will be used in emergency only.

CAUTION. A vessel may be aware of the existence of a practice area from PEXA Charts, Local Notices to Mariners or similar method of promulgation and by observing the warning signals or the practice.

The Range Authorities are responsible for ensuring that there should be no risk of damage from falling shell-splinters, bullets, etc., to any vessel which may be in a practice area.

If, however, a vessel finds herself in an area where practice is in progress, she should maintain her course and speed but, if she is prevented from doing this by the exigencies of navigation, it would assist the Range Authority if she would endeavour to clear the area at the earliest possible moment. Furthermore, if projectiles or splinters are observed to be falling near the vessel, all persons on board should take cover.

Fishermen operating in the vicinity of firing practice and exercise areas may occasionally bring unexploded missles or portions of them to the surface in their nets or trawls. These objects may be dangerous and should be treated with great circumspection and jettisoned immediately, no attempt being made to tamper with them or bring them back for inspection by Naval Authorities.

It is realised that the foregoing provisions do not apply in all respects in all countries. It is not, however, intended to repromulgate by Admiralty Notice information received about firing practice or exercise areas in foreign waters.

Areas are only in force intermittently or over limited periods, and local promulgation or warnings by radio, visual signals or Notices should be such that they will come to the attention of those whose co-operation or instruction is intended.

CHAPTER XV

EMERGENCY RADIOTELEGRAPH AND RADIOTELEPHONE PROCEDURES

1. On many radiotelegraph ships only one radio officer is carried. If by chance he were to be incapacitated through an accident, illness or other serious mishap whilst his ship was at sea, it might well be that there would be no one else on board capable of operating the radio equipment to send a distress call if one were necessary. It is clearly desirable that some provision should be made, to the extent that it is practicable, for one or more other officers on such ships to be capable of sending a distress call.

2. All ships which are fitted with radiotelegraph installations in compliance with the Merchant Shipping (Radio Installations) Regulations 1980 (as amended), and the Merchant Shipping (Radio) (Fishing Vessels) Rules, 1974 (as amended), are provided with an automatic keying device which, once it is set in operation, will first of all alert other ships by actuating their auto-alarms, and then transmit the distress call and also signals, repeated at intervals, which would be invaluable in enabling other ships to home on to the distressed ship by means of their direction-finding apparatus.

3. Investigation has shown that it is possible to produce a simple set of instructions which will enable the automatic keying device to be connected to the emergency transmitter and set in operation by an intelligent person unskilled in radio operation, provided that he can easily identify the controls which he needs to use. Moreover, the instructions need differ only slightly according to the types of equipment provided in the ship, the differences being in the description of the controls. A standard drill has been evolved, based on the operation of six controls on four pieces of equipment (the charging board, the aerial selector or switch unit, the emergency transmitter and the automatic keying device) and an outline of the instructions is given for information in the Appendix below. It is not intended for operational use.

4. A number of points arise from the use of the procedure outlined. In particular, it is emphasised that:—

(a) the marine radio companies have drawn up and can supply specific prodedures for operational use with each of their various types of equipment, based on the outline in the Appendix, and supplementing the control numbers with the precise description which is given to each control on each equipment;

(b) it is an essential partof the procedure that the controls in question should be prominently identified on the equipment by means of coloured (preferably yellow) labels numbered (preferably in red) to correspond with the operations referred to in the instructions in the Appendix. In cases where any confusion between individual controls might still be possible (e.g. in selecting the correct switch on an old-type charging board) the correct control knob or switch should be distinctively painted;

(c) to ensure the best possible response from the transmission of the distress call it is important for the person operating the equipment to wait, if possible, two minutes after the completion of the transmission of the alarm signal, before sending the distress call. This will give time for the radio officers on other ships, if off duty, to man their equipment after being alerted. Even one minute's delay is better than none at all;

(d) the procedure will be of little value in an emergency unless the ships' officers most likely to use it are practised in it, and the instructions applicable to their particular ship are ready to hand in a known place.

5. It is strongly recommended to owners and masters that, in radiotelegraph ships which carry only one radio officer, arrangements should be made for:—

(1) the detailed instructions referred to in paragraph 4 *(a)* of this Appendix to be posted conspicuously in the radiotelegraph room, preferably where they can be read by the light of the emergency lamp as well as by that of the main lighting system;

(2) yellow labels with red numbers to be affixed to the equipments, and for the correct controls to be suitably coloured wherever confusion might arise;

(3) deck officers subsequently to familiarise themselves with the procedure outined in the Appendix below and in the special instructions applicable to their ship.

6. The emergency conditions which might involve the use of this procedure might well also necessitate the use of the emergency lighting in the radiotelegraph room. Knowledge of the position of the door switch for the emergency lamp would be an important factor under these conditions. The door switch should be clearly labelled to indicate its purpose and the fact that the emergency light should only be used when the main source of light has failed.

APPENDIX

Outline of "INSTRUCTIONS TO ENABLE UNSKILLED PERSONS TO SEND A DISTRESS CALL IN AN EMERGENCY."

First ensure that the auto-alarm supply switch (AA) is in the "OFF" position. Then:—

A. On the charging board — Set battery switch to "DISCHARGE" position (by means of CONTROL No. 1).

B. On the Aerial selector or switch unit — Switch on the emergency transmitter (by means of CONTROL No. 3). *CHECK* whether transmitter is set for transmission on the DISTRESS frequency (500kHz), (and, if not, adjust appropriate controls as indicated in the detailed instructions posted in the radiotelegraph room).

D. On the automatic keying device — Connect to the emergency transmitter (by means of CONTROL No. 4). Set to "ALARM" (by means of CONTROL No. 5).

Start transmission (by means of CONTROL No. 6).

The ALARM SIGNAL, consisting of a series of 12 dashes, will now be sent out, the transmission taking one minute to complete. If circumstances permit, wait a further *two minutes* or as near two minutes as possible to allow the radio equipment of the ships which have been alerted through their auto-alarms to be manned, then:

Set automatic keying device to "DISTRESS" (by means of CONTROL No. 5).

Start transmission (by means of CONTROL No. 6).

A DISTRESS CALL, consisting of the international distress signal (SOS repeated three times), the morse characters for the word DE, the call sign of the ship repeated three times (if this facility has been provided in the automatic keying device), followed by a long dash or by two dashes each of 10 to 15 seconds' duration which will be used by other ships for direction-finding purposes, will now be sent out. If the equipment is left, the distress call will be repeated every twelve minutes until the battery is run down or the transmission is stopped (by setting CONTROL No. 4 at the "OFF" position) and the transmitter is switched off (by means of CONTROL No. 3). These repetitions will help searchers to fix the position of your ship and will provide radio beacon facilities for ships proceeding to your assistance.

NOTE This is only an outline of the procedure to be followed. It is quoted for information and should not be used operationally.

RADIOTELEPHONE ALARM SIGNAL
GENERATING DEVICE

The radiotelephone alarm signal, which consists of two tones transmitted alternately over a period of at least 30 seconds but not exceeding one minute, is intended primarily for use by ships in distress to give preliminary warning to other ships carrying radiotelephone equipment capable of receiving on the international radiotelephone distress frequency, 2,182 kHz, and to coast radio stations, of the impending transmission of a distress call and message by means of radiotelephony. Like the radiotelegraph alarm signal its use is permitted only for this purpose or to announce the loss of a person overboard in circumstances where the assistance of other ships is required and cannot be satisfactorily obtained by the use of the urgency signal only. The signal may be generated automatically by an electronic device, which is used in conjunction with a radiotelephone transmitter set to emit signals on 2,182 kHz.

United Kingdom coast radio stations have been transmitting the signals as a prelude to distress broadcasts on radiotelephony for several years, and so most radiotelephone operators will be aware of the alerting value of its distinctive warbling sound which can readily be recognized by ear through heavy interference.

The Merchant Shipping (Radio Installations) Regulations 1980, (as amended), and the Merchant Shipping (Radio) (Fishing Vessels) Rules, 1974 (as amended), require that ships which carry radiotelephone equipment in accordance with those Rules and Regulations shall carry the alarm signal generating device as part of the radio installation. As its use by a ship in distress will materially increase the ship's chances of being heard, and so of being helped, the Merchant Shipping (Radio Installations) Regulations (1980) (as amended) and the Merchant Shipping (Radio) (Fishing Vessels) Rules 1974 (as amended), require from 25th May 1982 onwards, that all United Kingdom passenger ships, cargo ships of 200 tons and upwards and fishing vessels of 12 metres or more in length carry one of these devices.

RADIOTELEPHONE DISTRESS PROCEDURE

1. Ships and fishing vessels compulsorily fitted with radiotelephone installations, in accordance with the Merchant Shipping (Radio Installations) Regulations 1980, (as amended), and the Merchant Shipping (Radio) (Fishing Vessels) Rules 1974, (as amended), are required to display in full view of the radiotelephone operating position a card or cards of instructions giving a clear summary of the radiotelephone distress, urgency and safety procedures.

2. Three cards are used. Two of these are instructional cards; one details the distress transmitting procedure, and the other includes the procedure to be followed on receipt of safety messages. The third and larger card includes details of the phonetic alphabet, figure-spelling table etc.

3. The form of cards to be displayed are given in this Appendix. The words printed in bold type should be printed in red.

4. A number of points arise from the use of the cards. In particular;

Card 1. The card should be so placed that it can easily be read from the radiotelephone operating position. Familiarity with the procedure will be greater if those concerned have the card before them at all times when they are on duty.

Card 2. Should be displayed at the place where the listening watch is maintained. Since the radio watch in the case of radiotelephony need not be kept by a qualified operator, there is a clear need for every person keeping such watch to be familiar with the prefixes "MAYDAY", "PAN PAN" and "SECURITE", and their meaning.

Card 3. Is primarily for use if language difficulties arise. There is no need for the card to be permanently displayed, but it is essential that those concerned know where to find such card when they are on duty.

5. Although provision of the cards is only mandatory for all radiotelephone ships and fishing vessels to which the Radio Rules and Radio Installations Regulations apply, it is strongly recommended that all ships voluntarily fitted with radiotelephone equipment using international or distress frequencies should also display the cards. The cards referred to in the above paragraphs are obtainable from the manufacturers of the radio equipment installed.

CHAPTER XVI

ABOUT FLAGS—RELATIVE TO THE ORIGIN OF THE UNION FLAG IN ITS PRESENT FORM—BRITISH NAVAL SIGNALLING FLAGS—INTERNATIONAL SALUTES—FLAGS TO BE WORN BY BRITISH MERCHANT SHIPS—FLAG ETIQUETTE

The Royal Standard

The Royal Standard is the personal flag of the Sovereign.

It is only hoisted on board a ship on occasions when the Sovereign is actually present.

Whenever the Sovereign shall go on board any ship of war the Royal Standard shall be hoisted at the main, the flag of the Lord High Admiral at the fore, and the Union Flag at the mizen of such ship; or, if on board a vessel with less than three masts, they shall be hoisted in the most conspicuous parts of her.

When Her Majesty the Queen, or the Duke of Edinburgh is embarked in any ship or vessel her or his standard shall be hoisted at the Main, and it shall be treated with the same respect and saluted in the same manner as the flags denoting the presence of the Sovereign.

Other Royal Standards

Personal standards are also appropriated for the use of:—
 (i) The Queen.
 (ii) The Queen Mother.
 (iii) The Duke of Edinburgh.
 (iv) Prince Charles.
 (v) Princess Margaret.
 (vi) Other Members holding the Title Royal Highness.

Admiral of the Fleet—The Union Flag is hoisted at the main by an Admiral of the Fleet as his proper flag.

The flag worn by an Admiral—A white flag with the red St. George's Cross thereon.

The flag worn by a Vice-Admiral—The same flag as above with one red ball in the upper canton of the flag next the staff.

The flag worn by a Rear-Admiral—The same flag as above, with one red ball in the upper canton and one in the lower canton next the staff.

The Broad Pendant of a Commodore

A Commodore of the first class shall wear a white broad pendant with a red St. George's Cross thereon.

A Commodore of the second class—The same flag as above, with a red ball in the upper canton of the broad pendant next the staff.

Senior Officers and Commission Pendants

When two or more of H.M. ships are present in a port or roadstead a small broad white pendant with the red St. George's Cross is to be hoisted by the senior officer's ships at the starboard topsail yardarm as a distinguishing flag in addition to the masthead pendant.

All H.M. ships in commission when not bearing a flag or broad pendant are to wear at one masthead a pendant as above.

RELATIVE TO THE ORIGIN OF THE UNION FLAG IN ITS PRESENT FORM

Union Flag

The original National Flag of England was the Banner of St. George (argent, a cross gules) to which the Banner of St. Andrew (azure, a saltire argent) was united, in pursuance of a Royal Proclamation, dated 12th April, 1606[1] of which the following is an extract:—

---

[1] NOTE—A Jack is a Flag to be flown only on the 'Jack' staff, i.e. a staff on the bowsprit or forepart of the ship. It is believed that the term 'Jack' is derived from the abbreviated name of the reigning Sovereign, King James I, under whose directon the flag was constructed and who signed his name 'Jacques'. Another derivation may have been from the 'Jack' or leather surcoat worn over the hauberk from the 14th to the 17th century inclusive, and which was emblazoned with the St. George's Cross. Jack is a contraction of 'Jazerine', a corruption of Ghaizerine (It.), a clinker-built boat, the Jack being formed of over-lapping plate of metal covered with cloth, velvet or leather.

In 1660, the Duke of York (afterwards James II) gave an order that the Union Flag should be worn only by the King's ships.

'Whereas some difference has arisen between Our Subjects of South and North Britain, travelling by seas, about the bearing of their flags: for the avoiding of all such contentions hereafter, We have, with the advice of Our Council, ordered that from henceforth, all Our subjects of this Isle and Kingdom of Great Britain, and the members thereof, shall bear in their maintops the Red Cross, commonly called St. George's Cross, and the White Cross, commonly called the St. Andrew's Cross, joined together, according to a form made by Our Heralds, and sent by Us to Our Admiral, to be published to our said subjects; and in their foretops Our subjects of South Britain shall wear the Red Cross only, as they were wont; and our subjects of North Britain in their foretops the White Cross only, as they were accustomed.'

Union with Scotland

On the 17th March, 1706-7, the Lords of the Committee of the Privy Council ordered the Kings of Arms and the Heralds to consider of the alterations to be made in the Ensigns Armorial, and the conjoining the Crosses of St. George and St. Andrew, to be used in all flags, banners, standards and ensigns at sea and on land.

On the 17th April, 1707, the Queen in Council, upon a report from the Lords of the Privy Council who were attended by the Kings of Arms and Heralds, with Divers drafts prepared by them, relating to the Ensigns Armorial for the United Kingdom, and conjoining the Crosses of St. George and St. Andrew, pursuant to the Act for Uniting the two Kingdoms, was pleased to approve of the following particulars (*inter alia*) 'that the Flags be according to the draft marked C, wherein the Crosses of St. George and St. Andrew are conjoined', as shown in the drawing entered in the College of Arms, with the Orders in Council.

Union with Ireland

On the 5th November, 1800, the King in Council was pleased to approve the report of a Committee of the Privy Council, that the Union Flag should be altered according to draft marked C, in which the Cross of St. George is conjoined with the Crosses of St. Andrew and St. Patrick, and which is thus described in the proclamation issued on the first day of January 1801:—

'And that the Union Flag shall be azure, the crosses saltires of St. Andrew and St. Patrick, quarterly per saltire counterchanged argent and gules; the latter fimbriated of the second surmounted by the Cross of St. George of the third fimbriated as the saltire.'

As to Red, White and Blue Ensigns

In 1687, an inquiry was made by Mr. Pepys, secretary to the Admiralty, as to the flags worn in the reign of Charles I; and in a manuscript in the British museum, it is stated that in the Duke of Buckingham's expedition to the Isle of Rhe in 1627, the Fleet was divided into the Red, Blue, and White Squadrons. The following is an extract from that manuscript:—

> 'The Duke, now lying at Portsmouth, divided his fleet into Squadrons. Himselfe Admirall and Generall in Chiefe went in ye Triumph bearing the Standard of England in ye maine topp, and Admirall particular of the bloody colours.
>
> 'The Earle of Lindsey was Vice Admirall, to the Fleete in the Rainbowe, bearing the King's usual colours in his fore topp, and a Blue Flag in his maine topp, and was Admirall of the blew colours.
>
> 'The Lord Harvey was rear Admirall in ye Repulse, bearing the King's usual colours in his mizen and a White Flag in the maine topp, and was Admirall of ye Squadron of white colours.'

There were two other squadrons, one under the Earl of Dandby with the St. George's flag, the other under Captain Pennington (made Admiral of his Squadron), with the St. Andrew's Cross.

It will be observed that in this instance the blue flag took precedence of the white.

In 1665, in the large fleet commanded by the Duke of York in person, as Lord High Admiral, Prince Rupert was Admiral of the White Squadron, and Sir Thomas Allen of the Blue; and an old drawing shows the three divisions of the fleet wearing ensigns of their respective colours.

But although the large fleets were thus divided, Admirals of foreign stations continued to wear the Union Flag at the main, fore, or mizen, according to their rank as full, vice, or rear-admiral.

Many instances are mentioned of the inconvenience arising from the use of the Royal Flag by private ships, and in 1660 the Duke of York gave an order that the Union Flag should be worn only be the King's ships.

It is clear that the sole object for which the three colours were formerly used was to distinguish the divisions of the fleet, which often numbered as many as 200 sail.

A variety of ensigns much increases the danger of confusion in action, and it may be observed that, in order to prevent that confusion, Lord Nelson, on going into action at Trafalgar, ordered the whole of his fleet to hoist the White Ensign; and it was under that flag, the 'Old Banner of England', but with the Union in the upper corner, that the victory was gained.

Latterly, owing to the comparatively small number of ships forming a fleet, the distinctive colours became of much less importance, while the frequent change of flags on foreign stations was very puzzling to foreigners, often led to mistakes, and in many ways was inconvenient; accordingly, on the 9th July, 1864, by Her Majesty's Order In Council of that date, it was directed that the classification of ships under the denominations of Red, White and Blue Squadrons should be discontinued, and that in future the 'White' Ensigns should be used by all of H.M. ships of war in commission; the Blue Ensigns by British merchant ships commanded by officers of the Royal Naval Reserve, after obtaining permission from the Admiralty; and the Red Ensign by all other ships and vessels belonging to H.M. subjects.

All other ships and vessels which belong to His Majesty's subjects shall wear a Red Ensign free from any badge or distinction mark, with the Union in the upper canton next the staff, except such yachts or vessels as may have warrants from the Admiralty to display other ensigns.

The following are the regulations as to the flags to be worn by any vessel maintained by any colony under the Colonial Defence Act of 1865:—

>   *(a)* Any vessel provided and used as a vessel of war shall wear the Blue Ensign with the badge of the colony in the fly thereof, and a Blue Pendant.
>
>   *(b)* All vessels belonging to or permanently in the service of the colonies, but not commissioned as vessels of war under the Act, shall wear a similar Blue Ensign but no Pendant.

Ships and vessels employed in the service of any public office shall carry a Blue Ensign, and a small Blue Flag with a Union described in a canton at the upper corner there of next to the staff, as a Jack: but in the centre of the fly of such Ensign and Jack, that is in the centre of that part between the Union and the end of the flag, shall be inserted the badge of the office to which they belong.

Hired transports are to wear the Blue Ensign with the Yellow Admiralty anchor in the fly; and when such vessels are in charge of commissioned officers of the Royal Navy, they are, in addition, to carry Blue Pennants with the Admiralty badge in the upper part next to the mast.

Hired vessels employed in the Surveying Service, when commanded by officers in Her Majesty's Navy, are to wear the Blue Ensign and Pendant.

BRITISH NAVAL SIGNALLING FLAGS

Measurement of Ensigns, Standards, etc.

The size of British Ensigns, Standards, and Union Flags is commonly expressed in terms of 'breadths'. This measurement which is 9 inches, was originally that of the width of a cloth of bunting, and it is still used for describing the size of any of the above flags, although bunting is now supplied in widths of 18 ins.

In the Union Flags, Standards, and the White, Blue, and Red Ensigns the length is twice the breadth. In the Admirals' Flags the length is 1½ times the breadth.

When the size of one of the above flags is spoken of as, say '16 breadths', it denotes that it is 16 × 9 inches (or 12 feet) broad and, as it is twice as long as it is broad, 24 feet long.

Each cloth of bunting (18 inches in width) has a few thicker threads worked not only into its edges, and also at every 6 inches of its warp,[2] and this mark shows the bunting to be of Government make.

Hoisting and Hauling Down the Ensign

1. H.M. ships when at anchor in home ports and roads shall hoist their ensigns at 8 o'clock in the morning from 25th March to 20th September inclusive, and at 9 o'clock from 21st September to 24th March inclusive; but when abroad, at 8 or 9 o'clock, as the Commander-in-Chief shall direct, and they shall be kept flying, if the weather permit, or the Senior Officer present see no objection thereto, throughout the day until sunset, when they are to be hauled down.

2. Whenever a ship shall come to anchor or get under way, if there be sufficient light for the ensign to be seen, it is to be hoisted, though earlier or later than aforesaid; also on her passing, meeting, joining, or parting from any other of H.M. ships and also, unless there should be sufficient reason to the contrary, on her falling in with any other ship or ships at sea, or when in sight of and near the land, and

---

[2] NOTE—The warp threads are those which are longitudinal to the cloth of bunting: the weft are those which are woven by the shuttle at right angles between them.

especially when passing or approaching forts, castles, batteries, lighthouses or towns.

INTERNATIONAL SALUTES

Naval

1. The Captain of a ship or the Senior Officer of more than one ship on anchoring at a foreign port where there is a fort or battery, or where a man-of-war of the nation may be lying, shall salute the National flag with 21 guns on being satisfied that the salute will be returned.

2. If one or more British ships of war meet a foreign ship of war bearing the flag of a Flag Officer senior in rank to the Senior Officer in command of H.M. ship or ships, such Senior Officer shall salute the foreign Flag Officer with the number of guns as laid down if in port after the proper national salutes shall have been interchanged.

3. A list of saluting stations is supplied to each ship.

4. The following regulations, in which the Maritime Powers generally have concurred, are observed in reference to the interchange of salutes between H.M. ships and foreign ships of war which bear the flag of a Flag Officer or the broad pendant of a Commodore, or a Captain commanding a Squadron or division:—

	Guns
The flag of an Admiral of the Fleet or Flag Officer who ranks with a Field-Marshal is to be saluted with	19
The flag of an Admiral	17
The flag of a Vice-Admiral	15
The flag of a Rear-Admiral	13
The broad pendant of a Commodore or a Captaine de Vaisseau Commandant de Division in the French Navy	11

As the rank of a full Admiral does not exist in the French Navy, Vice-Admirals of that nation whose flags may be hosited at the main are to be regarded as Full Admirals and are to be saluted with 17 guns.

5. Salutes to be returned:—

The following regulations are observed in regard to return salutes to and from H.M. ships and forts or batteries:—

(a) All salutes from foreign ships of war, either to H.M. ships or forts, are to be returned, gun for gun. Should there be no fort or battery from which such salutes can be returned, the Senior Naval Officer present will return them gun for gun.

(b) Salutes to the National Flag, on anchoring at a foreign port, are returned gun for gun.
  (c) Salutes to the flags of foreign Admirals and Commodores, when met with at sea, or in harbour, are returned gun for gun.
6. Salutes not to be returned:
  (a) Royal Salutes.
  (b) To H.M. subjects (except salutes to superior naval authorities).
  (c) To Royal personages, Chiefs of States, or members of Royal families, whether on arrival at, or departure from a port, or upon visiting ships of war.
  (d) To diplomatic, naval, military, or consular authorities, or to governors or officers administering a government, whether on arrival at, or departure from, a port, or when visiting ships of war.
  (e) To foreigners of high distinction on visiting ships of war.
  (f) Upon occasions of national festivals or anniversaries.

FLAGS TO BE FLOWN BY BRITISH MERCHANT SHIPS

1. THE RED ENSIGN.

Sections 73 and 74 of the Merchant Shipping Act, 1894, provide as follows:—

73—(1) The red ensign usually worn by Merchant Ships, without any defacement or modification whatsoever, is hereby declared to be the proper national colours for all ships and boats belonging to any British subject except in the case of Her Majesty's ships or boats, or in the case of any other ship or boat for the time being allowed to wear any other national colours in pursuance of a warrant from Her Majesty or from the Admiralty.

(2) If any distinctive national colours, except such red ensign, or except the Union Jack with a white border, or if any colours usually worn by Her Majesty's ships or resembling those of Her Majesty, or if the pendant usually carried by Her Majesty's ships, or any pendant resembling that pendant, are or is hoisted on board any ship or boat belonging to any British subject without warrant from Her Majesty or from the Admiralty, the master of the ship or boat or the owner thereof, if on board the same, and every other person hoisting the colours or pendant, shall for each offence incur a fine not exceeding five hundred pounds.

(3) Any commissioned officer on full pay in the military or naval

service of Her Majesty, or any officer of Customs in Her Majesty's dominions, or any British Consular officer, may board any ship or boat on which any colours or pendants are hoisted contrary to this Act, and seize and take away the colours or pendant and the colours or pendant shall be forfeited to Her Majesty.

(4) A fine under this Section may be removed with costs in the High Court in England or Ireland, or in the Court of Session in Scotland, or in any Colonial Court of Admiralty or Vice-Admiralty Court within Her Majesty's dominions.

(5) Any offence mentioned in this Section may also be prosecuted, and the fine for it recovered, summarily, provided that:
- *(a)* where any such offence is prosecuted summarily, the Court imposing the fine shall not impose a higher fine than one hundred pounds; and
- *(b)* nothing in this Section shall authorise the imposition of more than one fine in respect of the same offence.

74—(1) A ship belonging to a British subject shall hoist the proper national colours:
- *(a)* on a signal being made to her by one of Her Majesty's ships (including any vessel under the command of an officer of Her Majesty's Navy on full pay); and
- *(b)* on entering or leaving any foreign port; and
- *(c)* if of fifty tons gross tonnage or upwards, on entering or leaving any British port.

(2) If default is made on board any such ship in complying with this Section, the master of the ship shall for each offence be liable to a fine not exceeding one hundred pounds.

(3) This Section shall not apply to a fishing boat duly entered in the fishing boat register, and lettered and numbered as required by the Fourth Part of this Act.

2. THE BLUE ENSIGN.

(1) British Merchant Ships will be allowed to fly the Blue Ensign when the following conditions are fulfilled:—
- *(a)* The officer commanding the ship must be an officer on the Retired or Emergency List of the Royal Navy, or of the Royal Australian Navy, or an officer of the Royal Naval Reserve, of the Royal Australian Naval Reserve (Seagoing), of the Royal Canadian Naval Reserve, or of the Royal Naval Reserve (New Zealand Division).
- *(b)* The crew must include (in addition to the commanding officer) officers of the Royal Naval Reserve, of the Royal Australian Naval Reserve (Seagoing) of the Royal Canadian

Reserve, or of the Royal Naval Reserve (New Zealand Division), and men of the Royal Naval Reserve, of the Royal Australian Naval Reserve, of the Royal Candadian Naval Reserve, or of the New Zealand Royal Naval Reserve, Class B, to the number specified from time to time by the Admiralty, but officers on the Retired or Emergency List of the Royal Navy or of the Royal Australian Navy, men belonging to the Royal Fleet Reserve, to the Royal Australian Fleet Reserve, or to the New Zealand Royal Naval Reserve, Class A, Royal Naval pensioners and men holding Naval Reserved deferred pension certificates may be included in the number specified.

(c) Before hoisting the Blue Ensign the officer commanding the ship must be provided with an Admiralty warrant.

(d) The fact that the commanding officer holds a warrant authorising him to hoist the Blue Ensign must be noted on the Ship's Articles of Agreement.

(2) Commanding officers failing to fulfil the above conditions, unless such failure be due to death or other circumstances over which they have no control, will no longer be entitled to hoist the Blue Ensign.

(3) British Merchant Ships in receipt of Admiralty subvention will be allowed to fly the Blue Ensign, under Admiralty warrant.

(4) In order to ascertain that the above conditions are strictly carried out, the Captain of one of Her Majesty's ships meeting a ship carrying the Blue Ensign may send on board an officer not below the rank of Lieutenant, at any convenient opportunity, but this restriction as to the rank of the Boarding Officer is in no way to limit or otherwise affect the Authority or the duties of Naval officers either under the Merchant Shipping Acts or in time of war.

(5) Applications for permission to hoist the Blue Ensign on board British Merchant Ships in receipt of Admiralty subvention should be made direct to the Admiralty by the owners; for other Merchant Ships the applications should be made through the Registrar-General of Seamen.

Officers of the Naval Reserve who are desirous of obtaining the Admiralty warrant to fly the Blue Ensign on board the Merchant Ships should apply to the Registrar of Naval Reserve at any Mercantile Marine Office in the United Kingdom for a Form of Application (R.V.40), which, when complete, will be forwarded to the Registrar-General of Seamen.

FLAG ETIQUETTE
Law of Flags

The only National Ensign the use of which is free to all is the red. If, however, it has a special device it comes, like the White and Blue Ensigns, under the Admiralty control. The Red, White and Blue Ensign with special device may only be flown by those holding a personal Admiralty warrant. The use of the White Ensign is confined to the ships of the Royal Navy, Royal yachts, and yachts holding Admiralty warrants through the Royal Yacht Squadron.

Ensign (Wearing of)

The National Flag, i.e. ensign should be worn as follows:—
Steam vessels and other power craft.
  At anchor—At the ensign staff on the taffrail.
  Under way—At the ensign staff on the taffrail, or at the peak of the main gaff.
Sailing craft.
  At anchor—At the ensign staff on the taffrail.
  Under way—At the after peak.

The ensign should never be made up and broken out; moreover, it should be hoisted close up and not as is sometimes seen neither close up nor half mast with halyards slack.

When to fly—A ship belonging to a British subject shall hoist the proper national colours:—
  (a) On a signal being made to her by one of Her Majesty's ships (including any vessel under the command of an officer of Her Majesty's Navy on full pay); and
  (b) On entering or leaving any foreign port; and
  (c) If of 50 tons gross tonnage or upwards, on entering or leaving any British port.

When lying in port at home or abroad colours should be hoisted at 8 a.m. from 25th March to 20th September inclusive, and at 9 a.m. from 21st September to 24th March inclusive, and lowered at sunset. The time of sunset may be obtained from the Azimuth Tables.

When a vessel comes to anchor, or gets under way, before or after colours, she should fly her ensign providing there is sufficient light for it to be distinguished. Entering port under such circumstances, the colours should be lowered immediately after anchoring.

Dipping of Ensign

The dipping of the ensign is a salute and is carried out as follows:—

Slowly lower the ensign from the 'close up' to the 'dip', keeping the halyards taut, and when the salute has been acknowledged slowly hoist to the 'close up'.

When passing, or if at anchor being passed by Royal yachts or men-of-war ensigns should be dipped.

Position of Royal Standard when Hoisted on Merchant Ships

In the event of a visit by Her Majesty the Queen to a merchant ship, the appropriate place for the Royal Standard to be worn is at the mainmast head.

Dressing Ship

On ceremonial occasions, it is sometimes require to dress ship. This is carried out by hoisting masthead flags or running flags 'rainbow fashion'. In the latter case, on board a two-masted vessel, the line should run from stem to foremast head, thence to mainmast head and down to taffrail. On a single-masted vessel from stem to mainmast, thence to taffrail.

Particular attention is to be devoted to the arrangement of flags so that when flying they will have a symmetrical appearance. The flags should be evenly spaced and the same number of square flags between pendants. Since there are only two burgees in the International Code the best position for them is at each end of the line. The appearance of a string of flags is often completely spoilt by the line sagging in the middle. It is hard to avoid such an occurrance when flags are merely toggled together, but may be successfully overcome by fitting dressing lines. These lines may be made of light wire or small manila. The flags are seized to the lines, the middle of the hoist being secured to the line with a whipping of sailmaker's twine. When seizing the flags to the line the utmost care is to be taken to ensure the flags are evenly spaced; holidays between flags spoil the effect. If manila rope is used, be sure the rope is thoroughly stretched before setting up otherwise sagging is sure to occur.

Where a jib-boom is fitted, the effect is enhanced by running the dressing line from the jib-boom end and suspending flags from the jib-boom end and taffrail to the waterline. Affix weights to the ends of such lines to keep them as rigid as possible.

The order of the flags may be left to the discretion of the responsible officer. The ensign is carried in its proper place. No ensigns of any kind are to be used in dressing lines.

The flag to fly at the masthead depends upon the nature of the occasion for dressing ship. For instance, on the Queen's birthday fly the national ensign. When abroad and it is required to dress ship, in a single-masted vessel, masthead the ensign of the country visited, wearing, of course, the British ensign aft. In the case of a two-masted vessel fly at the foremast head the ensign of the country visited and masthead at the main the owner's House Flag.

A vessel should not be dressed 'rainbow fashion' when under way, but, instead, fly masthead ensign only.

Should you find on entering port all vessels at anchor dressed, endeavour to have your dressing lines all ready to hoist as soon as the anchor is let go.

Mourning

On days of national mourning, the ensign should be flown at half-mast. If abroad and the same circumstances prevail, out of respect the same procedure should be adopted.

To half-mast a flag, if not previously hoisted, it should be first hoisted 'close-up', then slowly lowered to half-mast. To lower a flag from half-mast, hoist to the 'close-up', then lower slowly.

If it is necessary to salute, or return a salute, when the ensign is at half mast, first hoist to the 'close-up', then 'dip' in the usual way. Keep the ensign at the 'close up' for a moment before lowering to half-mast.

CHAPTER XVII

INTERNATIONAL REGULATIONS FOR PREVENTING COLLISIONS AT SEA, 1983

PART A. GENERAL

Rule 1

Application

(a) These Rules shall apply to all vessels upon the high seas and in all waters connected therewith navigable by seagoing vessels.

(b) Nothing in these Rules shall interfere with the operation of special Rule made by an appropriate authority for roadsteads, harbours, rivers, lakes or inland waterways connected with the high seas and navigable by seagoing vessels. Such special Rules shall conform as closely as possible to these Rules.

(c) Nothing in these Rules shall interfere with the operation of any special Rules made by the Government of any State with respect to additional station or signal lights or whistle signals for ships of war and vessels proceeding under convoy, or with respect to additional station or signal lights for fishing vessels engaged in fishing as a fleet. These additional station or signal lights or whistle signals shall, so far as possible, be such that they cannot be mistaken for any light or signal authorized elsewhere under these Rules.

(d) Traffic separation schemes may be adopted by the Organization for the purpose of these Rules.

(e) Whenever the Government concerned shall have determined that a vessel of special construction or purpose cannot comply, fully with the provisions of any of these Rules with respect to the number, position, range or arc of visibility of lights or shapes, as well as to the disposition and characteristics of sound-signalling appliances, without interfering with the special function of the vessel, such vessel shall comply with such other provisions in regard to the number,

position, range or arc of visibility of lights or shapes, as well as to the disposition and characteristics of sound-signalling appliances, as her Government shall have determined to be the closest possible compliance with these Rules in respect to that vessel.

Rule 2

Responsibility

(a) Nothing in these Rules shall exonerate any vessel, or the owner, master or crew thereof, from the consequences of any neglect to comply with these Rules or of the neglect of any precaution which may be required by the ordinary practice of seamen, or by the special circumstances of the case.

(b) In construing and complying with these Rules due regard shall be had to all dangers of navigation and collision and to any special circumstances, including the limitations of the vessels involved, which may make a departure from these Rules necessary to avoid immediate danger.

Rule 3

General definitions

For the purpose of these Rules, except where the context otherwise requires:

(a) The word "vessel" includes every description of water craft, including non-displacement craft and seaplanes, used or capable of being used as a means of transportation on water.

(b) The term "power-driven vessel" means any vessel propelled by machinery.

(c) The term "sailing-vessel" means any vessel under sail provided that propelling machinery, if fitted, is not being used.

(d) The term "vessel engaged in fishing" means any vessel fishing with nets, lines, trawls or other fishing apparatus which restrict manoeuvrability, but does not include a vessel fishing with trolling lines or other fishing apparatus which do not restrict manoeuvrability.

(e) The word "seaplane" includes any aircraft designed to manoeuvre on the water.

(f) The term "vessel not under command" means a vessel which through some exceptional circumstance is unable to manoeuvre as required by these Rules and is therefore unable to keep out of the way of another vessel.

(g) The term "vessel restricted in her ability to manoeuvre" mean

a vessel which from the nature of her work is restricted in her ability to manoeuvre as required by these Rules and is therefore unable to keep out of the way of another vessel.

The following vessels shall be regarded as vessels restricted in their ability to manoeuvre:
  (i) a vessel engaged in laying, servicing or picking up a navigation mark, submarine cable or pipeline;
  (ii) a vessel engaged in dredging, surveying or underwater operations;
  (iii) a vessel engaged in replenishment or transferring persons, provisions or cargo while underway;
  (iv) a vessel engaged in the launching or recovery of aircraft;
  (v) a vessel engaged in minesweeping operations;
  (vi) a vessel engaged in a towing operation such as severely restricts the towing vessel and her tow in their ability to deviate from their course.

(h) The term "vessel constrained by her draught" means a power-driven vessel which because of her draught in relation to the available depth of water is severly restricted in her ability to deviate from the course she is following.

(i) The word "underway" means that a vessel is not at anchor, or made fast to the shore, or aground.

(j) The words "length" and "breadth" of a vessel mean her length overall and greatest breadth.

(k) Vessels shall be deemed to be in sight of one another only when one can be observed from the other.

(l) The term "restricted visibility" means any condition in which visibility is restricted by fog, mist, falling snow, heavy rainstorms, sandstorms or any other similar causes.

PART B. STEERING AND SAILING RULES

Section I. Conduct of vessels in any condition of visibility

Rule 4

Application

Rules in this Section apply in any condition of visibility.

Rule 5

Look-out

Every vessel shall at all times maintain a proper look-out by sight nd hearing as well as by all available means appropriate in the prevailing circumstances and conditions so as to make a full appraisal of the situation and of the risk of collision.

Rule 6

Safe speed

Every vessel shall at all times proceed at a safe speed so that she can take proper and effective action to avoid collision and be stopped within a distance appropriate to the prevailing circumstances and conditions.

In determining a safe speed the following factors shall be among those taken into account:

(a) By all vessels:
 (i) the state of visibility;
 (ii) the traffic density including concentration of fishing vessels or any other vessels;
 (iii) the manoeuvrability of the vessel with special reference to stopping distance and turning ability in the prevailing conditions;
 (iv) at night the presence of background light such as from shore lights or from back scatter of her own lights;
 (v) the state of wind, sea and current, and the proximity of navigation hazards;
 (vi) the draught in relation to the available depth of water.

(b) Additionally, by vessels with operational radar:
 (i) the characteristics, efficiency and limitations of the radar equipment;
 (ii) any constraints imposed by the radar range scale in use;
 (iii) the effect on radar detection of the sea state, weather and other sources of interference;
 (iv) the possibility that small vessels, ice and other floating objects may not be detected by radar at an adequate range;
 (v) the number, location and movement of vessels detected by radar;
 (vi) the more exact assessment of the visibility that may be possible when radar is used to determine the range of vessels or other objects in the vicinity.

Rule 7

Risk of collision

(a) Every vessel shall use all available means appropriate to the prevailing circumstances and conditions to determine if risk of collision exists. If there is any doubt such risk shall be deemed to exist

(b) Proper use shall be made of radar equipment if fitted and operational, including long-range scanning to obtain early warning of risk of collision and radar plotting or equivalent systematic observation of detected objects.

(c) Assumptions shall not be made on the basis of scanty information, especially scanty radar information.

(d) In determining if risk of collision exists the following considerations shall be among those taken into account:
> (i) such risk shall be deemed to exist if the compass bearing of an approaching vessel does not appreciably change.
> (ii) such risk may sometimes exist even when an appreciable bearing change is evident, particularly when approaching a very large vessel or a tow or when approaching a vessel at close range

Rule 8

Action to avoid collision

(a) Any action taken to avoid collision shall, if the circumstances of the case admit, be positive, made in ample time and with due regard to the observance of good seamanship.

(b) Any alteration of course and/or speed to avoid collision shall, if the circumstances of the case admit, be large enough to be readily apparent to another vessel observing visually or by radar; a succession of small alterations of course and/or speed should be avoided.

(c) If there is a sufficient sea room, alteration of course alone may be the most effective action to avoid a close-quarters situation provided that it is made in good time, is substantial and does not result in another close-quarters situation.

(d) Action taken to avoid collision with another vessel shall be such as to result in passing a safe distance. The effectiveness of the action shall be carefully checked until the other vessel is finally past and clear.

(e) If necessary to avoid collision or allow more time to assess the situation, a vessel shall slacken her speed or take all way off by stopping or reversing her means of propulsion.

Rule 9

Narrow channels

(a) A vessel proceeding along the course of a narrow channel or airway shall keep as near to the outer limit of the channel or fairway which lies on her starboard side as is safe and practicable.

(b) A vessel of less than 20 metres in length or a sailing vessel shall not impede the passage of a vessel which can safely navigate only within a narrow channel or fairway.

(c) A vessel engaged in fishing shall not impede the passage of any other vessel navigating within a narrow channel or fairway.

(d) A vessel shall not cross a narrow channel or fairway if such crossing impedes the passage of a vessel which can safely navigate only within such channel or fairway. The latter vessel may use the sound signal prescribed in Rule 34 *(d)* if in doubt as to the intention of the crossing vessel.

(e) (i) In a narrow channel or fairway when overtaking can take place only if the vessel to be overtaken has to take action to permit safe passing, the vessel intending to overtake shall indicate her intention by sounding the appropriate signal prescribed in Rule 34 *(c)* (i). The vessel to be overtaken shall, if in agreement, sound the appropriate signal prescribed in Rule 34 *(c)* (ii) and take steps to permit safe passing. If in doubt she may sound the signals prescribed in Rule 34 *(d)*.

(ii) This Rule does not relieve the overtaking vessel of her obligation under Rule 13.

(f) A vessel nearing a bend or an area of a narrow channel or fairway where other vessels may be obscured by an intervening obstruction shall navigate with particular alertness and caution and shall sound the appropriate signal prescribed in Rule 34 *(e)*.

(g) Any vessel shall, if the circumstances of the case admit, avoid anchoring in a narrow channel.

Rule 10

Traffic separation schemes

(a) This Rule applies to traffic separation schemes adopted by the Organization:

(b) A vessel using a traffic separation scheme shall:

(i) proceed in the appropriate traffic lane in the general direction of traffic flow for that lane;

(ii) so far as practicable keep clear of a traffic separation line or separation zone;

(iii) normally join or leave a traffic lane at the termination of the lane, but when joining or leaving from the side shall do so at as small an angle to the general direction of traffic flow as practicable.

(c) A vessel shall so far as practicable avoid crossing traffic lanes, but if obliged to do so shall cross as nearly as practicable at right angles to the general direction of traffic flow.

BROWN'S SIGNALLING

(d) Inshore traffic zones shall not normally be used by through traffic which can safely use the appropriate traffic lane within the adjacent traffic separation scheme.

(e) A vessel, other than a crossing vessel, shall not normally enter a separation zone or cross a separation line except:
  (i) in cases of emergency to avoid immediate danger;
  (ii) to engage in fishing within a separation zone.

(f) A vessel navigating in areas near the terminations of traffic separation schemes shall do so with particular caution.

(g) A vessel shall so far as practicable avoid anchoring in a traffic separation scheme or in areas near its terminations.

(h) A vessel not using a traffic separation scheme shall avoid it by as wide a margin as is practicable.

(i) A vessel engaged in fishing shall not impede the passage of any vessel following a traffic lane.

(j) A vessel of less than 20 metres in length or a sailing vessel shall not impede the safe passage of a power-driven vessel following a traffic lane.

(k) A vessel restricted in her ability to manoeuvre when engaged in an operation for the maintenance of safety of navigation in a traffic separation scheme is exempted from complying with this Rule to the extent necessary to carry out the operation.

(l) A vessel restricted in her ability to manoeuvre when engaged in an operation for the laying, servicing or picking up of a submarine cable, within a traffic separtion scheme, is exempted from complying with this Rule to the extent necessary to carry out the operation.

Section II. Conduct of vessels in sight of one another

Rule 11

Application

Rules in this Section apply to vessels in sight of one another.

Rule 12

Sailing vessels

(a) When two sailing vessels are approaching one another, so as to involve risk of collision, one of them shall keep out of the way of the other as follows:
  (i) when each has the wind on a different side, the vessel which has the wind on the port side shall keep out of the way of the other;
  (ii) when both have the wind on the same side, the vessel which is

to windward shall keep out of the way of the vessel which is to leeward;

(iii) if a vessel with the wind on the port side sees a vessel to windward and cannot determine with certainty whether the other vessel has the wind on the port or on the starboard side, she shall keep out of the way of the other.

(b) For the purposes of this Rule the windward side shall be deemed to be the side opposite to that on which the mainsail is carried or, in the case of a square-rigged vessel, the side opposite to that on which the largest fore-and-aft sail is carried.

Rule 13

Overtaking

(a) Notwithstanding anything contained in the Rules of this Section any vessel overtaking any other shall keep out of the way of the vessel being overtaken.

(b) A vessel shall be deemed to be overtaking when coming up with another vessel from a direction more than 22·5 degrees abaft her beam, that is, in such a position with reference to the vessel she is overtaking, that at night she would be able to see only the sternlight of that vessel but neither of her sidelights.

(c) When a vessel is in doubt as to whether she is overtaking another, she shall assume that this is the case and act accordingly.

(d) Any subsequent alteration of the bearing between the two vessels shall not make the overtaking vessel a crossing vessel within the meaning of these Rules or relieve her of the duty of keeping clear of the overtaken vessel until she is finally past and clear.

Rule 14

Head-on situation

(a) When two power-driven vessels are meeting on reciprocal or nearly reciprocal courses so as to invole risk of collision each shall alter her course to starboard so that each shall pass on the port side of the other.

(b) Such a situation shall be deemed to exist when a vessel sees the other ahead or nearly ahead and by night she could see the masthead lights of the other in a line or nearby in a line and/or both sidelights and by day she observes the corresponding aspect of the other vessel

(c) When a vessel is in any doubt as to whether such a situation exists she shall assume that it does exist and act accordingly.

Rule 15

Crossing situation

When two power-driven vessels are crossing so as to involve risk of collision, the vessel which has the other on her own starboard side shall keep out of the way and shall, if the circumstances of the case admit, avoid crossing ahead of the other vessel.

Rule 16

Action by give-way vessel

Every vessel which is directed to keep out of the way of another vessel shall, so far as possible, take early and substantial action to keep well clear.

Rule 17

Action by stand-on vessel

(a) (i) Where one of two vessels is to keep out of the way the other shall keep her course and speed.

(ii) The latter vessel may however take action to avoid collision by her manoeuvre alone, as soon as it becomes apparent to her that the vessel required to keepout of the way is not taking appropriate action in compliance with these Rules.

(b) When, from any cause, the vessel required to keep her course and speed finds herself so close that collision cannot be avoided by the action of the give-way vessel alone, she shall take such action as will best aid to avoid collision.

(c) A power-driven vessel which takes action in a crossing situation in accordance with sub-paragraph *(a)* (ii) of this Rule to avoid collision with another power-driven vessel shall, if the circumstances of the case admit, not alter course to port for a vessel on her own port side.

(d) This Rule does not relieve the give-way vessel of her obligation to keep out of the way.

Rule 18

Responsibilities between vessels

Except where Rules 9, 10 and 13 otherwise require:
(a) A power-driven vessel underway shall keep out of the way of:
(i) a vessel not under command;
(ii) a vessel restricted in her ability to manoeuvre;
(iii) a vessel engaged in fishing;

(iv) a sailing vessel.

(b) A sailing vessel underway shall keep out of the way of:
 (i) a vessel not under command;
 (ii) a vessel restricted in her ability to manoeuvre;
 (iii) a vessel engaged in fishing.

(c) A vessel engaged in fishing when underway shall, so far as possible keep out of the way of:
 (i) a vessel not under command;
 (ii) a vessel restricted in her ability to manoeuvre.

 (d) (i) any vessel other than a vessel not under command or a vessel restricted in her ability to manoeuvre shall, if the circumstances of the case admit, avoid impeding the safe passage of a vessel constrained by her draught, exhibiting the signals in Rule 28.
  (ii) a vessel constrained by her draught shall navigate with particular caution having full regard to her special condition.

(e) A seaplane on the water shall, in general keep well clear of all vessels and avoid impeding their navigation. In circumstances, however, where risk of collision exists, she shall comply with the Rules of this Part.

Section III. Conduct of vessels in restricted visibility

RULE 19

Conduct of vessels in restricted visibility

(a) This Rule applies to vessels not in sight of one another when navigating in or near an area of restricted visibility.

(b) Every vessel shall proceed at a safe speed adapted to the prevailing circumstances and conditions of restricted visibility. A power-driven vessel shall have her engines ready for immediate manoeuvre.

(c) Every vessel shall have due regard to the prevailing circumstances and conditions of restricted visibility when complying with the Rules of Section I of this Part.

(d) A vessel which detects by radar alone the presence of another vessel shall determine if a close-quarters situation is developing and/or risk of collision exists. If so, she shall take avoiding action in ample time, provided that when such action consists of an alteration of course, so far as possible the following shall be avoided:
 (i) an alteration of course to port for a vessel forward of the beam, other than for a vessel being overtaken;

(ii) an alteration of course towards a vessel abeam or abaft the beam.

(e) Except where it has been determined that a risk of collision does not exist, every vessel which hears apparently forward of her beam the fog signal of another vessel, or which cannot avoid a close-quarters situation with another vessel forward of her beam, shall reduce her speed to the minimum at which she can be kept on her course. She shall if necessary take all her way off and in any event navigate with extreme caution until danger of collision is over.

PART C. LIGHTS AND SHAPES

Rule 20

Application

(a) Rules in this Part shall be complied with in all weathers.

(b) The Rules concerning lights shall be complied with from sunset to sunrise, and during such times no other lights shall be exhibited, except such lights as cannot be mistaken for the lights specified in these Rules or do not impair their visibility or distinctive character, or interfere with the keeping of a proper look-out.

(c) the lights prescribed by these Rules shall, if carried, also be exhibited from sunrise to sunset in restricted visibility and may be exhibited in all other circumstances when it is deemed necessary.

(d) The Rules concerning shapes shall be complied with by day.

(e) The lights and shapes specified in these Rules shall comply with the provisions of Annex I to these Regulations.

Rule 21

Definitions

(a) "Masthead light" means a white light placed over the fore and aft centreline of the vessel showing an unbroken light over an arc of the horizon of 225 degrees and so fixed as to show the light from right ahead to 22·5 degrees abaft the beam on either side of the vessel.

(b) "Sidelights" means a green light on the starboard side and a red light on the port side each showing an unbroken light over an arc of the horizon of 112·5 degrees and so fixed as to show the light from right ahead to 22·5 degrees abaft the beam on its respective side. In a vessel of less than 20 metres in length the sidelights may be combined in one lantern carried on the fore and aft centreline of the vessel.

(c) "Sternlight" means a white light placed as nearly as practicable at the stern showing an unbroken light over an arc of the horizon of

135 degrees and so fixed as to show the light 67·5 degrees from right aft on each side of the vessel.

(d) "Towing Light" means a yellow light having the same characteristics as the "sternlight" defined in paragraph (c) of this Rule.

(e) "All round light" means a light showing an unbroken light over an arc of the horizon of 360 degrees.

(f) "Flashing light" means a light flashing at regular intervals at a frequency of 120 flashes or more per minute.

Rule 22

Visibility of lights

The lights prescribed in these Rules shall have an intensity as specified in Section 8 of Annex I to these Regulations so as to be visible at the following minimum ranges:

(a) In vessels of 50 metres or more in length:
—a masthead light, 6 miles
—a sidelight light, 3 miles;
—a sternlight, 3 miles;
—a towing light, 3 miles;
—a white, red, green or yellow all-round light, 3 miles.

(b) In vessels of 12 metres or more in length but less than 50 metres in length:
—a masthead light, 5 miles; except that where the length of the vessel is less than 20 metres, 3 miles;
—a sidelight, 2 miles;
—a sternlight, 2 miles;
—a towing light, 2 miles;
—a white, red, green or yellow all-round light, 2 miles.

(c) In vessels of less than 12 metres in length:
—a masthead light, 2 miles;
—a sidelight, 1 mile;
—a sternlight, 2 miles;
—a towing light, 2 miles;
—a white, red, green or yellow all-round light, 2 miles.

(d) In inconspicuous, partly submerged vessels or objects being towed:
—a white all-round light, 3 miles.

Rule 23

Power-driven vessels underway

(a) A power-driven vessel underway shall exhibit:
(i) a masthead light forward;

(ii) a second masthead light abaft of and higher than the forward one; except that a vessel of less than 50 metres in length shall not be obliged to exhibit such lights but may do so;
(iii) sidelights;
(iv) a sternlight.

(b) An air-cushion vessel when operating in the non-displacement mode shall, in addition to the lights prescribed in paragraph (a) of this Rule, exhibit an all-round flashing yellow light.

(c) (i) A power-driven vessel of less than 12 metres in length may in lieu of the lights prescribed in paragraph *(a)* of this Rule exhibit an all-round white light and sidelights;
(ii) a power-driven vessel of less than 7 metres in length whose maximum speed does not exceed 7 knots may in lieu of the lights prescribed in paragraph *(a)* of this Rule exhibit an all-round white light and shall, if practicable, also exhibit sidelights;
(iii) the masthead light or all-round white light on a power-driven vessel of less than 12 metres in length may be displaced from the fore and aft centreline of the vessel if centreline fitting is not practicable, provided that the sidelights are combined in one lantern which shall be carried on the fore and aft centreline of the vessel or located as nearly as practicable in the same fore and aft line as the masthead light or the all-round white light.

Rule 24

Towing and pushing

(a) A power-driven vessel when towing shall exhibit:
(i) instead of the light prescribed in Rule 23 *(a)* (i) or *(a)* (ii), two masthead lights forward in a vertical line. When the length of the tow, measuring from the stern of the towing vessel to the after end of the tow exceeds 200 metres, three such lights in a vertical line;
(ii) sidelights;
(iii) a sternlight;
(iv) a towing light in a vertical line above the sternlight;
(v) when the length of the tow exceeds 200 metres, a diamond shape where it can best be seen.

(b) When a pushing vessel land a vessel being pushed ahead are rigidly connected in a composite unit they shall be regarded as a power-driven vessel and exhibit the lights prescribed in Rule 23.

(c) A power-driven vessel when pushing ahead or towing alongside, except in the case of a composite unit, shall exhibit:

(i) instead of the light prescribed in Rule 23 (a) (i) or *(a)* (ii), two masthead lights forward in a vertical line;
(ii) sidelights;
(iii) a sternlight.

(d) A power-driven vessel to which paragraphs *(a)* and *(c)* of this Rule apply shall also comply with Rule 23 *(a)* (ii).

(e) A vessel or object being towed shall exhibit;
(i) sidelights;
(ii) a sternlight;
(iii) when the length of the tow exceeds 200 metres, a diamond shape where it can be best seen.

(f) Provided that any number of vessels being towed alongside or pushed in a group shall be lighted as one vessel,
(i) a vessel being pushed ahead, not being part of a composite unit, shall exhibit at the forward end, sidelights;
(ii) a vessel being towed alongside shall exhibit a sternlight and at the forward end, sidelights.

(g) An inconspicuous, partly submerged vessel or object, or combination of such vessels or objects being towed, shall exhibit:
(i) if it is less than 25 metres in breadth, one all-round white light at or near the forward end and one at or near the after end except that dracones need not exhibit a light at or near the forward end;
(ii) if it is 25 metres or more in breadth, two additional all-round white lights at or near the extremities of its breadth;
(iii) if it exceeds 100 metres in length, additional all-round white lights between the lights prescribed in sub-paragraphs (i) and (ii) so that the distance between the lights shall not exceed 100 metres;
(iv) a diamond shape at or near the aftermost extremity of the last vessel or object being towed and if the length of the tow exceeds 200 metres an additional diamond shape where it can best be seen and located as far forward as is practicable.

(h) Where from any sufficient cause it is impracticable for a vessel or object being towed to exhibit the lights or shapes prescribed in paragraph *(e)* or *(g)* of this Rule, all possible measures shall be taken to light the vessel or object towed or at least to indicate the presence of such vessels or object.

(i) Where from any sufficient cause it is impracticable for a vessel not normally engaged in towing operations to display the lights prescribed in paragraph *(a)* or *(c)* of this Rule, such vessel shall not be required to exhibit those lights when engaged in towing another

vessel in distress or otherwise in need of assistance. All possible measures shall be taken to indicate the nature of the relationship between the towing vessel and the vessel being towed as authorized by Rule 36, in particular by illuminating the towline.

RULE 25

Sailing vessels underway and vessels under oars

(a) A sailing vessel underway shall exhibit:
(i) sidelights;
(ii) a sternlight.

(b) In a sailing vessel of less than 20 metres in length the lights prescribed in paragraph *(a)* of this Rule may be combined in one lantern carried at or near the top of the mast where it can best be seen.

(c) A sailing vessel underway may in addition to the lights prescribed in paragraph *(a)* of this Rule, exhibit at or near the top of the mast, where they can best be seen, two all-round lights in a vertical line, the upper being red and the lower green, but these lights shall not be exhibited in conjunction with the combined lantern permitted by paragraph *(b)* of this Rule.

(d) (i) A sailing vessel of less than 7 metres in length shall, if practicable, exhibit the lights prescribed in paragraphs *(a)* or *(b)* of this Rule, but if she does not, she shall have ready at hand an electric torch or lighted lantern showing a white light which shall be exhibited in sufficient time to prevent collision.

(ii) A vessel under oars may exhibit the lights prescribed in this Rule for sailing vessels, but if she does not, she shall have ready at hand an electric torch or lighted lantern showing a white light which shall be exhibited in sufficient time to prevent collision.

(e) A vessel proceeding under sail when also being propelled by machinery shall exhibit forward where it can best be seen a conical shape, apex downwards.

RULE 26

Fishing vessels

(a) A vessel engaged in fishing, whether underway or at anchor, shall exhibit only the lights and shapes prescribed in this Rule.

(b) A vessel when engaged in trawling, by which is meant the dragging through the water of a dredge net or other apparatus used as a fishing appliance, shall exhibit:
(i) two all-round lights in a vertical line, the upper being green

and the lower white, or a shape consisting of two cones with their apexes together in a vertical line one above the other; a vessel of less than 20 metres in length may instead of this shape exhibit a basket;

(ii) a masthead light abaft of and higher than the all-round green light; a vessel of less than 50 metres in length shall not be obliged to exhibit such a light but may do so;

(iii) when making way through the water, in addition to the lights prescribed in this paragraph, sidelights and a sternlight.

(c) A vessel engaged in fishing, other than trawling, shall exhibit:

(i) two all-round lights in a vertical line, the upper being red and the lower white, or a shape consisting of two cones with apexes together in a vertical line one above the other; a vessel of less than 20 metres in length may instead of this shape exhibit a basket;

(ii) when there is outlying gear extending more than 150 metres horizontally from the vessel, an all-round white light or a cone apex upwards in the direction of the gear;

(iii) when making way through the water, in addition to the lights prescribed in this paragraph, sidelights and a sternlight.

(d) A vessel engaged in fishing in close proximity to other vessels engaged in fishing may exhibit the additional signals described in Annex II to these Regulations.

(e) A vessel when not engaged in fishing shall not exhibit the lights or shapes prescribed in this Rule, but only those prescribed for a vessel of her length.

Rule 27

Vessels not under command or restricted in their ability to manoeuvre

(a) A vessel not under command shall exhibit:

(i) two all-round red lights in a vertical line where they can best be seen;

(ii) two balls or similar shapes in a vertical line where they can best be seen;

(iii) when making way through the water, in addition to the lights prescribed in this paragraph, sidelights and a sternlight.

(b) A vessel restricted in her ability to manoeuvre, except a vessel engaged in minesweeping operations, shall exhibit:

(i) three all-round lights in a vertical line where they can best be seen. The highest and lowest of these lights shall be red and the middle light shall be white;

(ii) three shapes in a vertical line where they can best be seen. The highest and lowest of these shapes shall be balls and the middle one a diamond.;

(iii) when making way through the water, masthead lights, sidelights and a sternlight, in addition to the lights prescribed in sub-paragraph (i);

(iv) when at anchor, in addition to the lights or shapes prescribed in sub-paragraphs (i) and (ii), the light, lights or shape prescribed in Rule 30.

(c) A power-driven vessel engaged in towing operation such as severely restricts the towing vessel and her tow in their ability to deviate from their course shall, in addition to the lights or shapes prescribed in Rule 24 *(a)*, exhibit the lights or shapes prescribed in sub-paragraphs *(b)* (i) and (ii) of this Rule.

(d) A vessel engaged in dredging or underwater operations, when restricted in her abilty to manoeuvre, shall exhibit the lights and shapes prescribed in sub-paragraphs *(b)* (i), (ii) and (iii) of this Rule and shall in addition, when obstruction exists, exhibit:

(i) two all-round red lights or two balls in a vertical line to indicate the side on which the obstruction exists;

(ii) two all-round green lights or two diamonds in a vertical line to indicate the side on which another vessel may pass;

(iii) when at anchor, the lights or shaps prescribed in this paragraph instead of the lights or shape prescribed in Rule 30.

(e) Whenever the size of a vessel engaged in diving operations makes it impracticable to exhibit all lights and shapes prescribed in paragraph *(d)* of this Rule, the following shall be exhibited:

(i) three all-round lights in a vertical line where they can best be seen. The highest and lowest of these lights shall be red and the middle light shall be white;

(ii) a rigid replica of the International Code flag "A" not less than 1 metre in height. Measures shall be taken to ensure its all-round visibility.

(f) A vessel engaged in mineclearance operations shall in addition to the lights prescribed for a power-driven vessel in Rule 23 or to the lights or shape prescribed for a vessel at anchor in Rule 30 as appropriate, exhibit three all-round green lights or three balls. One of these lights or shapes shall be exhibited near the fore mast head and one at each end of the fore yard. These lights or shapes indicate that it is dangerous for another vessel to approach within 1000 metres of the mineclearance vessel.

168   BROWN'S SIGNALLING

(g) Vessels of less than 12 metres in length, except those engaged in diving operations, shall not be require to exhibit the lights and shapes prescribed in this Rule.

(h) The signals prescribed in this Rule are not signals of vessels in distress and requiring assistance. Such signals are contained in Annex IV to these Regulations.

RULE 28

Vessels constrained by their draught

A vessel constrained by her draught may, in addition to the lights prescribed for power-driven vessels in Rule 23, exhibit where they can best be seen three all-round red lights in a vertical line, or a cylinder.

RULE 29

Pilot vessels

(a) A vessel engaged on pilotage duty shall exhibit:
  (i) at or near the masthead, two all-round lights in a vertical line, the upper being white and the lower red;
 (ii) when underway, in addition, sidelights and a sternlight;
(iii) when at anchor, in addition to the lights prescribed in sub-paragraph (i), the light, lights or shape prescribed in Rule 30 for vessels at anchor.

(b) A pilot vessel when not engaged on pilotage duty shall exhibit the lights or shapes prescribed for a similar vessel of her length.

RULE 30

Anchored vessels and vessels aground

(a) A vessel at anchor shall exhibit where it can best be seen:
  (i) in the fore part, an all-round white light or one ball;
 (ii) at or near the stern and at a lower level than the light prescribed in sub-paragraph (i), an all-round white light.

(b) A vessel of less than 50 metres in length may exhibit an all-round white light where it can best be seen instead of the lights prescribed in paragraph *(a)* of this Rule.

(c) A vessel at anchor may, and a vessel of 100 metres and more in length shall, also use the available working or eqivalent lights to illuminate her decks.

(d) A vessel aground shall exhibit the lights prescribed in

paragrpahs *(a)* or *(b)* of this Rule and in addition, where they can best be seen:
  (i) two all-round red lights in a vertical line;
  (ii) three balls in a vertical line.

(e) A vessel of less than 7 metres in length, when at anchor or aground, not in or near a narrow channel, fairway or anchorage, or where other vessels normally navigate, shall not be required to exhibit the lights or shapes prescribed in paragraphs *(a),* and *(b)* of this Rule.

Rule 31

Seaplanes

Where it is inpracticable for a seaplane to exhibit lights and shapes of the characteristics or in the positions prescribed in the Rules of this Part she shall exhibit lights and shapes as closely similar in characteristics and position as is possible.

PART D. SOUND AND LIGHT SIGNALS

Rule 32

Definitions

(a) The word "whistle" means any sound signalling appliance capable of producing the prescribed blasts and which complies with the specifications in Annex III to these Regulations.

(b) The term "short blasts" means a blast of about one second's duration.

(c) The term "prolonged blast" means a blast of from four to six seconds' duration.

Rule 33

Equipment for sound signals

(a) A vessel of 12 metres or more in length shall be provided with a whistle and a bell and a vessel of 100 metres or more in length shall, in addition, be provided with a gong, the tone and sound of which cannot be confused with that of a bell. The whistle, bell and gong shall comply with the specifications in Annex III to these Regulations. The bell or gong or both may be replaced by other equipment having the same respective sound characteristics, provided that manual sounding of the required signals shall always be possible.

(b) A vessel of less than 12 metres in length shall not be obliged to

carry the sound signalling appliances prescribed in paragraph *(a)* of this Rule but is she does not, she shall be provided with some other means of making an efficient sound signal.

Rule 34

Manoeuvring and warning signals

(a) When vessels are in sight of one another, a power-driven vessel underway, when manoeuvring as authorized or required by these Rules, shall indicate that manoeuvre by the following signals on her whistle:
—one short blast to mean "I am altering my course to starboard";
—two short blast to mean "I am altering my course to port";
—three short blasts to mean "I am operating astern propulsion".

(b) Any vessel may supplement the whistle signals prescribed in paragraph *(a)* of this Rule by light signals, repeated as appropriate, whilst the manoeuvre is being carried out:
 (i) these light signals shall have the following significance:
    —one flash to mean "I am altering my course to starboard";
    —two flashes to mean "I am altering my course to port";
    —three flashes to mean "I am operating astern propulsion";
 (ii) the duration of each flash shall be about one second, the interval between flashes shall be about one second, and the interval between successive signals shall be not less than ten seconds;
 (iii) the light used for this signal shall, if fitted, be an all-round white light, visible at a minimum range of 5 miles, and shall comply with the provisions of Annex I.

(c) When in sight of one another in a narrow channel or fairway:
 (i) a vessel intending to overtake another shall in compliance with Rule 9 *(e)* (i) indicate her intention by the following signals on her whistle:
    —two prolonged blasts followed by one short blast to mean "I intend to overtake you on your starboard side";
    —two prolonged blasts followed by two short blasts to mean "I intend to overtake you on your port side";
 (ii) the vessel about to be overtaken when acting in accordance with Rule 9 *(e)* (i) shall indicate her agreement by the following signal on her whistle:
    —one prolonged, one short, one prolonged and one short blast, in that order.

(d) When vessels in sight of one another are approaching each other and from any cause either vessel fails to understand the

intentions or actions of the other or is in doubt whether sufficient action is being taken by the other to avoid collision, the vessel in doubt shall immediately indicate such doubt by giving at least five short and rapid blasts on the whistle. Such signal may be supplemented by a light signal of at least five short and rapid flashes.

(e) A vessel nearing a bend or an area of a channel or fairway where other vessels may be obscured by an intervening obstruction shall sound one prolonged blast. Such signal shall be answered with a prolonged blast by any approaching vessel that may be within hearing around the bend or behind the intervening obstruction.

(f) If whistles are fitted on a vessel at a distance apart of more than 100 metres, one whistle only shall be used for giving manoeuvring and warning signals.

Rule 35

Sound signals in restricted visibility

In or near an area of restricted visibility, whether by day or night, the signals prescribed in this Rule shall be used as follows:

- *(a)* A power-driven vessel making way through the water shall sound at intervals of not more than 2 minutes one prolonged blast.
- *(b)* A power-driven vessel underway but stopped and making no way through the water shall sound at intervals of not more than 2 minutes two prolonged blasts in succession with an interval of about 2 seconds between them.
- *(c)* A vessel not under command, a vessel restricted in her ability to manoeuvre, a vessel constrained by her draught, a sailing vessel, a vessel engaged in fishing, and a vessel engaged in towing or pushing another vessel shall, instead of the signals prescribed in paragraphs *(a)* or *(b)* of this Rule, sound at intervals of not more than 2 minutes three blasts in succession, namely one prolonged followed by two short blasts.
- *(d)* A vessel engaed in fishing, when at anchor, and a vessel restricted in her ability to manoeuvre when carrying out her work at anchor, shall instead of the signals prescribed in paragraph *(g)* of this Rule sound the signal prescribed in paragraph *(c)* of this Rule.
- *(e)* A vessel towed or if more than one vessel is towed the last vessel of the tow, if manned, shall at intervals of not more than 2 minutes sound four blasts in succession, namely one prolonged followed by three short blasts. When practicable, this signal shall be made immediately after the signal made by the towing vessel.
- *(f)* When a pushing vessel and a vessel being pushed ahead are

rigidly connected in a composite unit they shall be regarded as a power-driven vessel and shall give the signals prescribed in paragraphs *(a)* or *(b)* of this Rule.

(g) A vessel at anchor shall at intervals of not more than one minute ring the bell rapidly for about 5 seconds. In a vessel of 100 metres or more in length the bell shall be sounded in the forepart of the vessel and immediately after the ringing of the bell the gong shall be sounded rapidly for about 5 seconds in the after part of the vessel. A vessel at anchor may in addition sound three blasts in succession, namely one short, one prolonged and one short blast, to give warning of her position and of the possibility of collision to an approaching vessel.

(h) A vessel aground shall give the bell signal and if required the gong signal prescribed in paragraph *(g)* of this Rule and shall, in addition, give three separate and distinct strokes on the bell immediately before and after the rapid ringing of the bell. A vessel aground may in addition sound an appropriate whistle signal.

(i) A vessel of less than 12 metres in length shall not be obliged to give the above-mentioned signals but, if she does not, shall make some other efficient sound signal at intervals of not more than 2 minutes.

(j) A pilot vessel when engaged on pilotage duty may in addition to the signals prescribed in paragraphs *(a)*, *(b)* or *(g)* of this Rule sound an identity signal consisting of four short blasts.

Rule 36

Signals to attract attention

If necessary to attract the attention of another vessel any vessel may make light or sound signals that cannot be mistaken for any signal authorized elsewhere in these Rules, or may direct the beam of her searchlight in the direction of the danger, in such a way as not to embarrass any vessel.

Any light to attract the attention of another vessel shall be such that it cannot be mistaken for any aid to navigation. For the purpose of this Rule the use of high intensity intermittent or revolving lights, such as strobe lights, shall be avoided.

Rule 37

Distress signals

When a vessel is in distress and requires assistance she shall use or exhibit the signals prescribed in Annex IV to these Regulations.

PART E. EXEMPTIONS

Rule 38

Exemptions

Any vessel (or class of vessels) provided that she complies with the requirements of the International Regulations for Preventing Collisions at Sea, 1960, the keel of which laid or which is at a corresponding stage of construction before the entry into force of these Regulations may be exempted from compliance therewith as follows:

(a) The installation of lights with ranges prescribed in Rule 22, until four years after the date of entry into force of these Regulations.

(b) The installation of lights with colour specifications as prescribed in Section 7 of Annex I to these Regulations, until four years after the date of entry into force of these Regulations.

(c) The repositioning of lights as a result of conversion from Imperial to metric units and rounding off measurement figures, permanent exemption.

(d) (i) The repositioning of masthead lights on vessels of less than 150 metres in length, resulting from the prescriptions of Section 3 *(a)* of Annex I, permanent exemption.

(ii) The repositioning of masthead lights on vessels of 150 metres or more in length, resulting from the prescriptions of Section 3 *(a)* of Annex I to these Regulations, until nine years after the date of entry into force of these Regulations.

(e) The repositioning of masthead lights resulting from the prescriptions of Section 2 *(b)* of Annex I, until nine years after the date of entry into force of these Regulations.

(f) The repositioning of sidelights resulting from the prescriptions of Sections 2 *(g)* and 3 *(b)* of Annex I, until nine years after the date of entry into force of these Regulations.

(g) The requirements for sound signal appliances prescribed in Annex III, until nine years after the date of entry into force of these Regulations.

(h) The repositioning of all-round lights reulting from the prescription of Section 9 *(b)* of Annex I to these Regulations, permanent exemption.

ANNEX I

Positioning and technical details of lights and shapes

1. *Definition*

The term "height above the hull" means height above the uppermost continuous deck. This height shall be measured from the position vertically beneath the location of the light.

2. *Vertical positioning and spacing of lights*

(a) On a power-driven vessel of 20 metres or more in length the masthead lights shall be placed as follows:
  (i) the forward masthead light, or if only one masthead light is carried, then that light, at a height above the hull of not less than 6 metres, and, if the breadth of the vessel exceeds 6 metres, then at a height above the hull not less than such breadth, so however that the light need not be placed at a greater height above the hull than 12 metres;
  (ii) when two masthead lights are carried the after one shall be at least 4·5 metres vertically higher than the forward one.

(b) The vertical separation of masthead lights of power-driven vessels shall be such that in all normal conditions of trim the afterlight will be seen over and separate from the forward light at a distance of 1,000 metres from the stem when viewed from sea level.

(c) The masthead light of a power-driven vessel of 12 metres but less than 20 metres in length shall be placed at a height above the gunwale of not less than 2·5 metres.

(d) A power-driven vessel of less than 12 metres in length may carry the uppermost light at a height of less than 2·5 metres above the gunwale. When however a masthead light is carried in addition to sidelights and a sternlight, then such masthead light shall be carried at least 1 metre higher than the sidelights.

(e) One of the two or three masthead lights prescribed for a power-driven vessel when engaged in towing or pushing another vessel shall be placed in the same position as either the forward masthead light or the after masthead light; provided that, if carried on the aftermast, the lowest after masthead light shall be at least 4·5 metres vertically higher than the forward masthead light.

(f) (i) The masthead light or lights prescribed in Rule 23 *(a)* shall be so placed as to be above and clear of all other lights and obstructions except as described in sub-paragraph (ii).

(ii) When it is impracticable to carry the all-round lights prescribed by Rule 27 *(b)* (i) or Rule 28 below the masthead lights, they may be carried above the after masthead light(s) or vertically in between the forward masthead light(s) and after masthead light(s), provided that in the latter case the requirement of Section 3 *(c)* of this Annex shall be complied with.

(g) The sidelights of a power-driven vessel shall be placed at a height above the hull not greater than three-quarters of that of the forward masthead light. They shall not be so low as to be interfered with by deck lights.

(h) The sidelights, if in a combined lantern and carried on a power-driven vessel of less than 20 metres in length, shall be placed not less than 1 metre below the masthead light.

(i) When the Rules prescribe two or three lights to be carried in a vertical line, they shall be spaced as follows:
  (i) on a vessel of 20 metres in length or more such lights shall be spaced not less than 20 metres apart, and the lowest of these lights shall, except where a towing light is required, be placed at a height of not less than 4 metres above the hull.
  (ii) on a vessel of less than 20 metres in length such lights shall be spaced not less than 1 metre apart and the lowest of these lights shall, except where a towing light is required, be placed at a height of not less than 2 metres above the hull.
  (iii) when three lights are carried they shall be equally spaced.

(j) The lower of the two all-round lights prescribed for a vessel when engaged in fishing shall be at a height above the sidelights not less than twice the distance between the two vertical lights.

(k) The forward anchor light prescribed in Rule 30 *(a)* (i), when two are carried, shall not be less than 4·5 metres above the after one. On a vessel of 50 metres or more in length this forward anchor light shall be placed at a height of not less than 6 metres above the hull.

3. *Horizontal positioning and spacing of lights*

(a) When two masthead lights are prescribed for a power-driven vessel, the horizontal distance between them shall not be less than one-half of the length of the vessel but need not be more than 100 metres. The forward light shall be placed not more than one-quarter of the length of the vessel from the stem.

(b) On a power-driven vessel of 20 metres or more in length the sidelights shall not be placed in front of the forward masthead lights. They shall be placed at or near the side of the vessel.

(c) When the lights prescribed in Rule 27 *(b)* (i) or Rule 28 are placed vertically between the forward masthead light(s) and the after masthead light(s) these all-round lights shall be placed at a horizontal distance of not less than 2 metres from the fore and aft centreline of the vessel in the athwartship direction.

4. *Details of location of direction-indicating lights for fishing vessels, dredgers and vessels engaged in underwater operations*

(a) The light indicating the direction of the outlying gear from a vessel engaged in fishing as prescribed in Rule 26 *(c)* (ii) shall be placed at a horizontal distance of not less than 2 metres and not more than 6 metres away from the two all-round red and white lights. This light shall be placed not higher than the all-round white light prescribed in Rule 26 *(c)* (i) and not lower than the sidelights.

(b) The lights and shapes on a vessel engaged in dredging or underwater operations to indicate the obstructed side and/or the side on which it is safe to pass, as prescribed in Rule 27 *(d)* (i) and (ii), shall be placed at the maximum practical horizontal distance, but in no case less than 2 metres, from the lights or shapes prescribed in Rule 27 *(b)* (i) and (ii). In no case shall the upper of these lights or shapes be at a greater height than the lower of the three lights or shapes prescribed in Rule 27 *(b)* (i) and (ii).

5. *Screens for sidelights*

The sidelights of vessels of 20 metres or more in length shall be fitted with inboard screens painted matt black, and meeting the requirements of Section 9 of this Annex. On vessels of less than 20 metres in length the sidelights, if necessary to meet the requirements of Section 9 of this Annex, shall be fitted with inboard matt black screens. With a combined lantern, using a single vertical filament and a very narrow division between the green and red sections, external screens need not be fitted.

6. *Shapes*

   (a) Shapes shall be black and of the following sizes:
   (i) a ball shall have a diameter of not less than 0·6 metre;
   (ii) a cone shall have a base diameter of not less than 0·6 metre and a height equal to its diameter;
   (iii) a cylinder shall have a diameter of at least 0·6 metre and a height of twice its diameter;
   (iv) a diamond shape shall consist of two cones as defined in (ii) above having a common base.

   (b) The vertical distance between shapes shall be at least 1·5 metres.

(c) In a vessel of less than 20 metres in length shapes of lesser dimensions but commensurate with the size of the vessel may be used and the distance apart may be correspondingly reduced.

7. *Colour specification of lights*

The chromaticity of all navigation lights shall conform to the following standards, which lie within the boundaries of the area of the diagram specified for each colour by the International Commission on Illumination (CIE).

The boundaries of the area for each colour are given by indicating the corner co-ordinates, which are as follows:

(i) *White*

| x | 0·525 | 0·525 | 0·452 | 0·310 | 0·310 | 0·443 |
| y | 0·382 | 0·440 | 0·443 | 0·348 | 0·283 | 0·382 |

(ii) *Green*

| x | 0·028 | 0·009 | 0·300 | 0·203 |
| y | 0·385 | 0·723 | 0·511 | 0·356 |

(iii) *Red*

| x | 0·680 | 0·660 | 0·735 | 0·721 |
| y | 0·320 | 0·320 | 0·265 | 0·259 |

(iv) *Yellow*

| x | 0·612 | 0·618 | 0·575 | 0·575 |
| y | 0·382 | 0·382 | 0·425 | 0·406 |

8. *Intensity of lights*

(a) The minimum luminous intensity of lights shall be calculated by using the formula:

$$I = 3·43 \times 10^6 \times T \times D^2 \times K^{-D}$$

where I is luminous intensity in candelas under service conditions,

T is threshold factor 2×10^{-7} lux,

D is range of visibility (luminous range) of the light in nautical miles,

K is atmospheric transmissivity.

For prescribed lights the value of K shall be 0·8, corresponding to a meteorological visibility of approximately 13 nautical miles.

(b) A selection of figures derived from the formula is given in the following table:

Range of visibility (*luminous range*) of light in nautical miles D	Luminous intensity of light in candelas for $K = 0.8$ I
1	0·9
2	4·3
3	12
4	27
5	52
6	94

Note: the maximum luminous intensity of navigation lights should be limited to avoid undue glare. This shall not be achieved by a variable control of the luminous intensity.

9. *Horizontal sectors*

(a) (i) In the forward direction, sidelights as fitted on the vessel shall show the minimum required intensities. The intensities must decrease to reach practical cut-off between 1 degree and 3 degrees outside the prescribed sectors.

(ii) For sternlights and masthead lights and at 22·5 degrees abaft the beam for sidelights, the minimum required intensities shall be maintained over the arc of the horizon up to 5 degrees within the limits of the sectors prescribed in Rule 21. From 5 degrees within the prescribed sectors the intensity may decrease by 50 per cent up to the prescribed limits; it shall decrease steadily to reach practical cut-off at no more than 5 degrees outside the prescribed sectors.

(b) All-round lights shall be so located as not to be obscured by masts, topmasts or structures within angular sectors of more than 6 degrees, except anchor lights prescribed in Rule 30, which need not be placed at an impracticable height above the hull.

10. *Vertical sectors*

(a) The vertical sectors of electric lights as fitted, with the exception of lights on sailing vessels shall ensure that:
  (i) at least the required minimum intensity is maintained at all angles from 5 degrees above to 5 degrees below the horizontal;
  (ii) at least 60 per cent of the required minimum intensity is maintained from 7·5 degrees above to 7·5 degrees below the horizontal.

(b) In the case of sailing vessels the vertical sectors of electric lights as fitted shall ensure that:

(i) at least the required minimum intensity is maintained at all angles from 5 degrees above to 5 degrees below the horizontal;
(ii) at least 50 per cent of the required minimum intensity is maintained from 25 degrees above to 25 degrees below the horizontal.

(c) In the case of lights other than electric these specifications shall be met as closely as possible.

11. *Intensity of non-electric lights*

Non-electric lights shall so far as practicable comply with the minimum intensities, as specified in the Table given in Section 8 of this Annex.

12. *Manoeuvring light*

Notwithstanding the provisions of paragraph 2 *(f)* of this Annex the manoeuvring light described in Rule 34 *(b)* shall be placed in the same fore and aft vertical plane as the masthead light or lights and, where practicable, at a minimum height of 2 metres vertically above the forward masthead light, provided that it shall be carried not less thn 2 metres vertically above or below the after masthead light. On a vessel where only one masthead light is carried, the manoeuvring light, if fitted, shall be carried where it can best be seen, not less than 2 metres vertically apart from the masthead light.

13. *Approval*

The construction of lights and shapes and the installation of lights on board the vessel shall be to the satisfaction of the appropriate authority of the State whose flag the vessel if entitled to fly.

ANNEX II

Additional signals for fishing vessels fishing in close proximity

1. *General*

The lights mentioned herein shall, if exhibited in pursuance of Rule 26 *(d)*. be placed where they can best be seen. They shall be at least 0·9 metres apart but at a lower level than the lights prescribed in Rule 26 *(b)* (i) and *(c)* (i). The lights shall be visible all round the horizon at a distance of at least 1 mile but at a lesser distance than the lights prescribed by these Rules for fishing vessels.

2. *Signals for trawlers*

(a) Vessels when engaged in trawling, whether using demersal or pelagic gear, may exhibit:
(i) when shooting their nets:

two white lights in a vertical line;
(ii) when hauling their nets:
one white light over one red light in a vertical line;
(iii) when the net has come fast upon an obstruction:
two red lights in a vertical line.

(b) Each vessel engaged in pair trawling may exhibit:
(i) by night, a searchlight directed forward and in the direction of the other vessel of the pair;
(ii) when shooting or hauling their nets or when their nets have come fast upon an obstruction, the lights prescribed in 2 *(a)* above.

3. *Signals for purse seiners*

Vessels engaged in fishng with purse seine gear may exhibit two yellow lights in a vertical line. These lights shall flash alternately every second and with equal light and occultation duration. These lights may be exhibited only when the vessel is hampered by its fishing gear.

ANNEX III

Technical details of sound signal appliances

1. *Whistles*

(a) Frequencies and range of audibility

The fundamental frequency of the signal shall lie within the range 70-700 Hz.

The range of audibility of the signal from a whistle shall be determined by those frequencies, which may include the fundamental and/or one or more higher frequencies, which lie within the range 180-700 Hz ($\pm$ 1 per cent) and which provide the sound pressure levels specified in paragraph 1 *(c)* below.

(b) Limits of fundamental frequencies

To ensure a wide variety of whistle characteristics, the fundamental frequency of a whistle shall be between the following limits:
(i) 70-200 Hz, for a vessel 200 metres or more in length;
(ii) 130-350 Hz, for a vessel 75 metres but less than 200 metres in length;
(iii) 250-700 Hz, for a vessel less than 75 metres in length.

(c) Sound signal intensity and range of audibility

A whistle fitted in a vessel shall provide, in the direction of maximum intensity of the whistle and at a distance of 1 metre from it, a sound pressure level in at least one 1/3rd-octave band within the range of frequencies 180-700 Hz ($\pm$ 1 per cent) of not less than the appropriate figure given in the table below.

Length of vessel in metres	1/3rd-octave band level at 1 metre in dB referred to 2×10^{-5} N/m^2	Audibility range in nautical miles
200 or more...	143	2
75 but less than 200...	138	1·5
20 but less than 75...	130	1
Less than 20...	120	0·5

The range of audibility in the table above is for information and is approximately the range at which a whistle may be heard on its forward axis with 90 per cent probablility in conditions of still air on board a vessel having average background noise level at the listening posts (taken to be 86 dB in the octave band centred on 250 Hz and 63 dB in the octave band centred on 500 Hz).

In practice the range at which a whistle may be heard is extremely variable and depends critically on weather conditions; the values given can be regarded as typical but under conditions of strong wind or high ambient noise level at the listening post the range may be much reduced.

(d) Directional properties

The sound pressure level of a directional whistle shall be not more than 4 dB below the prescribed sound pressure level on the axis at any direction in the horizontal plane within $\pm$ 45 degrees of the axis. The sound pressure level at any other direction in the horizontal plane shall be not more than 10 dB below the prescribed sound pressure level on the axis, so that the range in any direction will be at least half the range on the forward axis. The sound pressure level shall be measured in that 1/3rd-octave band which determines the audibility range.

(e) Positioning of whistles

When a directional whistle is to be used as the only whistle on a vessel, it shall be installed with its maximum intensity directed straight ahead.

A whistle shall be placed as high as practicable on a vessel, in order to reduce interception of the emitted sound by obstructions and also to minimize hearing damage risk to personnel. The sound pressure level of the vessel's own signal at listening posts shall not exceed 110 dB (a) and so far as practicable should not exceed 100 dB (A).

(f) Fitting of more than one whistle

If whistles are fitted at a distance apart of more than 100 metres, it shall be so arranged that they are not sounded simultaneously.

(g) Combined whistle systems

If due to the presence of obstructions the sound field of a single whistle or of one of the whistles referred to in paragraph 1 *(f)* above is liklely to have a zone of greatly reduced signal level, it is recommended that a combined whistle system be fitted so as to overcome this reduction. For the purposes of the Rules a combined whistle system it to be regarded as a single whistle. The whistles of a combined system shall be located at a distance apart of not more than 100 metres and arranged to be sounded simultaneously. The frequency of any one whistle shall differ from those of the others by at least 10 Hz.

2. *Bell or gong*

(a) Intensity of signal

A bell or gong, or other device having similar sound characteristics shall produce a sound pressure level of not less than 110 dB at a distance of 1 metre from it.

(b) Construction

Bells and gongs shall be made of corrosion-resistant material and designed to give a clear tone. The diameter of the mouth of the bell shall be not less than 300 mm for vessels of 20 metres or more in length, and shall be not less than 200 mm for vessels of 12 metres or more but of less than 20 metres in length. Where practicable, a power-driven bell striker is recommended to ensure constant force but manual operation shall be possible. The mass of the striker shall be not less than 3 per cent of the mass of the bell.

3. *Approval*

The construction of sound signal appliances, their performance and their installation on board the vessel shall be to the satisfaction of the appropriate authority of the State whose flag the vessel is entitled to fly.

ANNEX IV

Distress signals

1. The following signals, used or exhibited either together or separately, indicated distress and need of assistance.

- (a) A gun or other explosive signal fired at intervals of about a minute;
- (b) a continuous sounding with any fog-signalling apparatus;
- (c) rockets or shells, throwing red stars fixed one at a time at short intervals;
- (d) a signal made by radiotelegraphy or by any other signalling method consisting of the group ▪ ▪ ▪ ━ ━ ━ ▪ ▪ ▪ (SOS) in the Morse Code;
- (e) a signal sent by radiotelephony consisting of the spoken word "Mayday";
- (f) the International Code Signal of distress indicated by N.C.;
- (g) a signal consisting of a square flag having above or below it a ball or anything resembling a ball;
- (h) flames on the vessel (as from a burning tar barrel, oil barrel, etc.);
- (i) a rocket parachute flare or a hand flare showing a red light;
- (j) a smoke signal giving off orange-coloured smoke;
- (k) slowly and repeatedly raising and lowering arms outstretched to each side;
- (l) the radiotelegraph alarm signal;
- (m) the radiotelephone alarm signal;
- (n) signals transmitted by emergency position-indicating radio beacons.

2. The use of exhibition of any of the foregoing signals except for the purpose of indicating distress and need of assistance and the use of other signals which may be confused with any of the above signals is prohibited.

3. Attention is drawn to the relevant sections of the International Code of Signals, the Merchant Ship Search and Rescue Manual and the following signals:
- (a) a piece of orange-coloured canvas with either a black square and circle or other appropriate symbol (for identification from the air);
- (b) a dye marker.

CHAPTER XVIII

GENERAL NOTICES

All candidates for Department of Transport Certificates are required to be conversant with the general information, e.g. Notices regarding distress signals, life-saving, various special signals, etc., contained in The Annual Summary of Admiralty Notices to Mariners for the information of Masters of Foreign-going Ships, Home Trade Ships and Fishing Vessels, which is published in January each year.

In addition, for Foreign-going vessels, the following Notices to Mariners are issued by the Admiralty:—*(a)* Daily Notices *(b)* Weekly (complete) edition of Notices, *(c)* a Quarterly edition. For Home Trade and Fishing Vessels, the following Notices to Mariners are issued:—*(a)* Daily Notices, *(b)* Weekly (home-trade) edition of Notices.

As the information given in these Notices is liable to be altered or cancelled, or new information given, candidates are strongly advised to procure the latest issues, which should be carefully studied, prior to presenting themselves for examination.

The Notices to Mariners may be obtained from any Mercantile Marine Office in the United Kingdom, free of charge.

The following is important information culled from Notices to Mariners:—

Signals used in connection with the Life-Saving Services on the Coast of the United Kingdom

In the event of your ship being in distress off, or stranded on, the coast of the United Kingdom, the following signals shall be used by life-saving stations when communicating with your ship, and by your ship when communicating with life-saving stations.

(a) Replies from life-saving stations or maritime rescue units to distress signals made by a ship or person:—

Signals	Signification
By day—Orange smoke signal or combined lights and sound signal (thunderlight) consisting of three single signals which are fired at intervals of approximately one minute............	"You are seen—assistance will be given as soon as possible."
By night—White star rocket consisting of three single signals which are fired at intervals of approximately one minute.....	(Repetition of such signals shall have the same meaning.)

If necessary the day signals may be given at night or the night signals by day.

(b) Landing signals for the guidance of small boats with crews or persons in distress:—

Signals	Signification
By day.—Vertical motion of a white flag or the arms or signalling the code letter "K"(— ‧ —) given by light or sound-signal apparatus...................	
By night.—Vertical motion of a white light or flare, or signalling the code letter "K" (— ‧ —) given by light or sound-signal apparatus. A range (indication of direction) may be given by placing a steady white light or flare at a lower level and in line with the observer............	"This is the best place to land."
By day.—Horizontal motion of a white flag or arms extended horizontally or signalling the code letter "S"(‧ ‧ ‧) given by light or sound-signal apparatus	"Landing here highly dangerous"
By night.—Horizontal motion of a white light or flare or signalling the code letter "S" (‧ ‧ ‧) given by light or sound-signal apparatus..............	

Co-operation between a ship's crew and H.M. Coastguard in the use of rocket rescue equipment.

Should lives be in danger and your vessel be in a position where rescue by the rocket rescue equipment is possible, a rocket with line attached will be fired from the shore across your vessel. Get hold of this line as soon as you can. When you have got hold of it, signal to the shore as indicated in paragraph 43 *(c)*.

Should your vessel carry a line-throwing appliance, it may be preferable to use this and fire a line ashore, but this should not be done without first consulting the rescue company on shore. If this method is used, the rocket line may not be of sufficient strength to haul out the whip and jackstay and those on shore will secure it to a stouter rocket line. When this is done, they will signal as indicated in paragraph 43 *(c)*. On seeing the signal, haul in the line which was fired from the vessel until the stouter line is on board.

Then, when the rocket line is held, make the appropriate signal to the shore (paragraph 43 *(c)*) and proceed as follows:—

(1) When you see the appropriate signal i.e. "haul away", made from the shore, haul upon the rocket line until you get a tail block with an endless fall rove through it (called the "whip"), and with a jackstay attached to the becket of the tail block.

(2) Cut or cast off the rocket line and make the tail block fast, close up to the mast or other convenient position, bearing in mind that the fall should be kept clear from chafing any part of the vessel. Before cutting or casting off the rocket line, make sure that you have the tail attached to the block well in hand. When the tail block is made fast, signal to the shore again (as in paragraph 43 *(c)*).

(3) As soon as this signal is seen, the shore party will then set the jackstay taut, and by means of the whip, will haul the breeches buoy out to the ship. The person to be rescued should get in to the breeches buoy and sit well down. When he is secure he should signal again to the shore as indicated in paragraph 43 *(c)* and the men on shore will haul the person in the breeches buoy to the shore. When he is landed the empty breeches buoy will be hauled back to the ship. This operation will be repeated until all persons are landed.

(4) During the course of the operations should it be necessary to signal, either from your ship to the shore, or from the shore to your ship, to "Slack away" or "Avast hauling" this should be done as indicated in paragraph 43 *(c)*.

It may sometimes happen that the state of the weather and/or the condition or position of the ship will require the aforementioned procedures to be modified. Where this is the case, the rescue company will always attempt to advise you of the procedures to be followed.

All Coastguard rescue companies are equipped with VHF radios and the rescue operation as a whole will be greatly facilitated if communication with the rescue company is established on VHF Channel 16, as soon as possible. In the absence of radio communication with the rescue company, the system of signalling indicated in paragraph 43 *(c)* must be strictly followed. However if communication by flashing light is necessary, a large majority of rescue companies have trained signalmen.

Normally, all women, children, passengers and helpless persons should be landed before the crew of the vessel but there may be occasions when, perhaps because of communications difficulties between the casualty and the rescue company ashore, it would be sensible if the first person to be landed were a responsible member of the ship's crew.

A poster drawing attention to the provisions of paragraphs 44-52 can be obtained on application to the Central Form Store, Department of Environment, Victoria Road, South Ruislip, Middx., HA4 0NZ.

Use of rocket-line throwing apparatus between ships

Where an assisting ship proposes to establish communication by means of a line-throwing apparatus she should before making her final approach ascertain whether or not it is safe for her to fire the rocket, particularly if the other ship is a tanker. If it is safe she should manoeuvre to WINDWARD before firing over the other ship's deck. If not, she should go to LEEWARD and prepare to receive a line. EXTREME CAUTION must be exercised when firing line-throwing rockets between ships when helicopters are in the vicinity.

When a vessel in distress is carrying petrol spirit or other highly inflammable liquid and is leaking, the following signals should be exhibited to show that it is dangerous to fire a line-carrying rocket by reason of the risk of ignition:—

By day.—Flag B of the International Code of Signals hoisted at the masthead.

By night.—A red light hoisted at the masthead.

When visibility is bad the above signals should be supplemented by the use of the following International Code signal made in sound:—

GU (— — · · · —) "It is not safe to fire a rocket."

Use of aircraft in assisting ships

R.A.F. aircraft (other than helicopters) used on SAR duties usually carry droppable survival equipment and pyrotechnics. These aircraft may be able to assist a ship in distress by:—
  (1) Locating her when her position is in doubt and informing the shore authorities so that ships in the vicinity going to her assistance may be given her precise position;
  (2) Guiding surface craft to the casualty or, if the ship has been abandoned, to survivors in lifeboats, on rafts or in the sea;
  (3) Keeping the casualty under observation;
  (4) Making a position by marine marker, smoke float or flame float and illuminating an area to assist rescue operations;
  (5) Dropping survival equipment.
  Helicopters may be able to pick up survivors (see paragraph 58) but their carrying capacity is limited.

INFORMATION CONCERNING SUBMARINES
PART 1—WARNING SIGNALS

1. Mariners are warned that considerable hazard to life may result from the disregard of the following signals, which denote the presence of submarines:—

(a) **Visual signals**

British vessels fly the international code Group "NE2" to denote that submarines, which may be submerged, are in the vicinity. Vessels are cautioned to steer so as to give a wide berth to any vessel flying this signal. If from any cause it is necessary to approach her a good look-out must be kept for submarines whose presence may be indicated only by their periscopes or snorts showing above the water.

A submarine submerged at a depth too great to show her periscope may occasionally indicate her position by red-and-white or red-and-yellow buffs or floats, which tow on the surface close astern.

(b) **Pyrotechnics and Smoke Candles**

The following signals are used by submerged submarines:—

Signal	*Signification.*
White smoke candle (with or without flame) (may be accompanied by yellowish green fluorescent dye)........ Yellow smoke candles....... Green pyro flares launched approx. 50 feet into the air and burning for about 10 sec.	Indicates position in response to request from ship or aircraft or as required.

Red pyro flares or red smoke. (may be accompanied by other smoke candles repeated as often as possible).........	Keep clear. I am carrying out emergency surfacing procedure. Do not stop propellers. Clear area immediately. Stand by to render assistance.
Alternating red flare/white smoke	Submarine in distress. Take action as above.

Note:—If the red pyro flare signal is sighted and the submarine does not surface within 5 minutes it should be assumed that the submarine is in distress and has sunk. An immediate attempt should be made to fix the position in which the signal was sighted, after which action in accordance with Part III should be taken.

Two white or yellow smoke candles released singly about 3 minutes apart.....................	Keep clear I am preparing to surface. Do not stop propellers. Clear the immediate vicinity.

2. It must not be inferred from the above that submarines exercise only when in company with escorting vessels.

3. The notice "Submarine Exercise Area" on certain charts should not be read to mean that submarines do not exercise outside such areas. Under certain circumstances warnings that submarines are exercising in specified areas may be broadcast by a British Telecom coast radio station.

PART II—NAVIGATION LIGHTS

4. Submarines may be met on the surface by night, particularly in the following areas and their approaches:—

Portsmouth, Portland, Plymouth, Barrow, North Channel, Clyde and Forth Areas.

5. The masthead and side lights of H.M. Submarines are placed well forward and very low over the water in proportion to the length and tonnage of these vessels. In particular, the forward masthead light may be lower than the side lights and the after masthead light may be well forward of the mid-point of the submarine's length. Stern lights are placed very low indeed and may at times be partially obscured by spray and wash. They are invariably lower than the side

lights. While at anchor or at buoy by night submarines display an all round white light amidships in addition to the normal anchor lights. The after anchor light of nuclear submarines may be mounted on the upper rudder which is some distance astern of the hull's surface waterline. Care must be taken to avoid confusion with two separate vessels of less than 50 metres in length.

6. The overall arrangement of submarine's lights are therefore unusual and may well give the impression of markedly smaller and shorter vessels than they are. Their vulnerability to collision when proceeding on the surface and the fact that some submarines are nuclear powered dictates particular caution when approaching them. Some submarines are fitted with a yellow (amber) *quick-flashing* light situated about 1 to 2 metres above the after masthead light. This additional light is for use as an aid to identification in narrow water and areas of dense traffic. The rate of flash of the submarine fitted light is 90 flashes per minute (except Denmark—105 flashes per minute): this should not be confused with a similar light used by Hovercraft currently with a rate of 120 flashes per minute.

Certain submarines of the Royal Navy are fitted with quick-flashing amber anti-collision lights. These lights flash at between 90 and 105 flashes per minute and, due to the configurations of the various classes, are fitted 1 to 2 metres above or below the masthead light.

The showing of one of these quick-flashing lights is intended to indicate to an approaching vessel the need for added caution rather than to give immediate identification of the type of vessel exhibiting such lights. Subsequent identification of submarine or hovercraft can usually be made by observation.

PART III—SUNKEN SUBMARINE

7. A bottomed submarine which is unable to surface will try to indicate her position by the following methods:—

(a) Releasing an indicator buoy (which carries a vertical whip aerial) as soon as the accident occurs (see Figs. 1 & 2).

(b) On the approach of surface vessels and at regular intervals by firing candles giving off yellow, red or white smoke. As far as possible yellow candles or the red/white pyrotechnic will be used by day.

Note:—It should be noted that it may be impossible for a submarine to fire her smoke candles. correspondingly a partially flooded submarine may have only a certain number of her smoke candles available and searching ships should not therefore expect many to appear.

Since oil slicks or debris may be the only indication of the presence or whereabouts of the sunken submarine, it it vitally important that surface ships refrain from discharging anything which might appear to have come from a submarine while they are in the submarine probability area. Searching ships and aircraft can waste many valuable hours investigating these false contacts.

Some submarine pyrotechnics can be fitted with message carriers. If a message has been attached, the pyrotechnic will be fitted with a dye marker, giving off a yellowish-green dye on the surface. Such a pyrotechnic should be recovered as soon as it has finished burning.

(c) Pumping out oil fuel or lubricating oil.

(d) Blowing out air.

8. British and some Allied submarines are fitted with two Indicator Buoys one each end of the ship, which can be released from inside in case of emergency or if for any reason, the submarine is unable to surface. A description of the Indicator Buoys is given in paragraph 17.

9. In any submarine accident time is the most vital factor affecting the chances of rescue of survivors, and as the sighting of an indicator buoy may be the first intimation that an accident has in fact occurred, it is vital that no time should be lost in taking action.

10. The sighting of any buoy answering the attached description should at once be reported by the quickest available means to the Navy, coastguard, or police. However, if vessels are unable to establish communications without leaving the vicinity of the submarine, it should be borne in mind that the primary considertion should be for vessels to remain standing by to rescue survivors and not leave the scene of the accident. Every effort should be made to include in the report the serial number of the buoy; this number is affixed below the word "Foreward" or "Aft", as shown in the attached plates.

11. Indicator buoys are attached to the submarine by a length of wire. The 0050 and 0060 buoys have 915 metres and the 0700 buoy 1,830 metres of wire. These buoys may therefore still be secured to a submarine in depths of 500 fathoms and 1,000 fathoms respectively. In areas where strong tidal streams or currents are prevalent the depth from which the buoy may be expected to watch is considerably reduced and in these areas it is possible that a buoy may only watch at slack tide. It is possible that indicator buoys may break adrift accidentally even though the parent submarine may not have sunk, similarly a buoy found to be adrift is not necessarily an indication that all is well since it may have broken adrift after being deliberately

released following an accident. In any case it is therefore important to establish by the most seamanlike practicable means whether or not the buoy is adrift. The 0060 and 0070 buoy wire is 0·4 inch circumference fibre core galvanised steel wire rope with a nominal breaking strain of 1,000 lbs. Its total weight in water is 2lb per 100 ft. length. Although, if no other means is available, the lowering of a boat and the weighing of the wire by hand is permissable very great care should be exercised in this operation since it is absolutely vital not to part the wire. In no circumstances should the boat be secured to the buoy or turns taken on the wire once it has been established that the latter is not adrift.

12. At any time after a submarine accident survivors may start attempting to escape. Current policy dictates that survivors will wait before escaping until:—

  (a) Rescue vessels are known to be standing-by.

or *(b)* conditions inside the submarine deteriorate to such an extent that an attempt to escape must be made.

It should be noted that, in certain circumstances, situation *(b)* may not arise through lack of air supply until a time after the accident of several days. However, if the submarine is badly damaged, survivors may have to make an escape attempt immediately. Any ship finding a moored submarine indicator buoy should not therefore leave the position but should stand by well clear ready to pick up survivors. The latter will ascend nearly vertically, and it is plainly important plenty of sea room is given to enable them to do so in safety. On arrival on the surface men may be exhausted or ill, and if circumstances are favourable therefore the presence of a boat already lowered is very desirable. Some men may require a recompression chamber, and it will therefore be the aim of the Naval authorities to get such a chamber to the scene as soon as possible.

13. In order that those trapped in the submarine shall be made aware that help is at hand Naval vessels drop small charges into the sea which can be heard from inside the submarine. There is no objection to the use of small charges for this purpose; but it is vital that they are not dropped too close since men in the process of making ascents are particularly vulnerable to under water explosions, and may easily receive fatal injuries. A distance of a quarter of a mile is considered to be safe. If no small charges are available, the running of an echo sounder or the banging of the outer skin of the ship's hull with a hammer from a position below the water-line is likely to be heard in the submarine, and such banging and/or sounding should therefore be carried out at frequent intervals.

14. Submarines may at any time release pyrotechnic floats which on reaching the surface burn with flame and/or smoke thus serving to mark the position of the wreck. They are likely to acknowledger sound signals by this means.

15. To sum up, the aims of a submarine rescue operations are:—

(a) To fix the exact position of the submarine.

(b) To get a ship standing by to pick up survivors, if practicable with boats already lowered.

(c) To get medical assistance to survivors picked up.

(d) To get a diver's recompression chamber to the scene in case this is required by those seriously ill after being exposed to great pressure.

(e) To inform the trapped men that help is at hand.

16. There is a large Naval organization designed to fulfil these aims, which is always kept at instant readiness for action. It is clear, however, that any ship may at any time find evidence of a submarine disaster, and if she takes prompt and correct action as described above she may be in a position to play a vital part.

17. Description of Submarine Indicator Buoys

Two types of submarine indicator buoy are in existence:—

Cylindrical in shape and made of aluminium 2ft. 3 inches in diameter and 18½ inches deep, and a cylindrical projection on the bottom about 6 inches deep. On the side are two fittings which carry a stirrup, from which is suspended a length of ½ inch circumference steel mooring wire. The buoy floats end up with a freeboard of about 6" at slack water.

It is painted with "International Orange" and has a ring carrying 'cats eyes' around the light.

For Identification purposes the buoys are allocated a serial number (e.g. 043). The serial number is affixed under the word 'Forward' or 'Aft'. Also inscribed around the top of the buoys are the words:—FINDER INFORM NAVY, COASTGUARD OR POLICE. DO NOT SECURE TO OR TOUCH.

A light which flashes approximately once every second, over a period of about 60 hours, is mounted in the centre of the top surface. In darkness, and during good weather the visibility of the light without binoculars is 3,500 yards.

The buoy carries a whip aerial approximately 5' 6" tall and is fitted with an automatic transmitting radio unit operating on 4,340 kHz.

The signal is transmitted automatically when the indicator buoy is released and is as follows:—

Transmission.	Number of Repetitions.	Duration.
3 figure serial number	3 times	30 seconds
S.O.S.	6 times	30 "
SUBSUNK	3 times	30 "
Long Mark	Once	30 "

The whole message will be made twice through giving a total transmitting time of 4 minutes. There will be a silence for the next 6 minutes. Thereafter the complete 10 minute cycle will be repeated. The transmission should continue for a maximum of 36 hours.

Note:—Ships hearing these signals should report to the Navy or Coastguard giving their position and, if possible, an indication of signal strength.

Semi-spherical in shape and made of low density expanded plastic covered with glassfibre reinforced plastic skin of $1/8''$ thick for physical protection, 2' 6" in diameter. Anchorage for 3 mm diameter fibre core type A galvanised steel mooring wire at bottom of buoy slightly offset and has reflective tapes in the form of longitudinal strips provided, alternate red and white.

For identification purposes the buoys are allocated a serial number (e.g. 043). The serial number is affixed under the word "Forward" or "Aft". Also inscribed around the top of the buoy are the words FINDER INFORM NAVY, COASTGUARD OR POLICE: DO NOT SECURE TO OR TOUCH.

A light which flashes approximately every 2 seconds, over a period of about 72 hours, is mounted in the centre of the top surface. In darkness and during good weather, the unassisted visibility of the light is 5 nautical miles.

The buoy carries HF and UHF whip aerials (168 cm and 100 cm long respectively) and is fitted with two automatic transmitting radio units which operate on 8364 kHz and 243·0 MHz.

The transmitters are automatically activate when the indicator buoy is released and their transmission sequences are as follows:—

HF Transmitter

	Transmissions	Number of Repetitions	Duration
(1)	3 figure serial number	3 times	21—30s
(2)	120 (±5)s silent period		
(3)	3 figure serial number	3 times	21—30s
(4)	SOS	6 times	27s
(5)	SUBSUNK	3 times	36s
(6)	Long D/F mark	Once	30s

The message will then be repeated from (3) giving a total transmitting sequence of 6 minutes duration. There will then be a silent period for 2 minutes before the whole sequence re-commences. The transmissions will continue for a minimum of 72 hours.

UHF Transmitter

The UHF emission will consist of 3 audio sweeps from 1600 Hz down to not lower than 300 Hz, occupying a period of 1·2s. The emission will then be silent for 2·8s. The transmission duration should continue for a minimum of 72 hours.

18. The accompanying plates show submarine indicator buoys, smoke candles fired from submarines, sonobuoys, aircraft float, smoke and flame, and markers.

White Smoke Candles. These are fired from submarines to indicate their position. They burn for up to 10 minutes emitting white smoke and flame and can thus be seen by day or night; they can easily be confused with aircraft marine markers and floats, smoke and flame. The candle can also give off a yellowish-green dye indicating that a message is attached.

Yellow Smoke Candles. These are fired from submarines to indicate their position. They burn for about 5 minutes emitting yellow smoke. They can be seen more easily than the white smoke candles in rough weather but cannot be seen at night.

Red and White Pyrotechnic. These are fired to indicate the position of a submarine which is unable to surface. They burn for about 8 minutes emitting alternate red flares and white smoke. A fluorescent dye marker and message carrier are attached to the buoy which is maintained in an upright position on the surface by means of 3 petal floats.

Sonobuoys. These are dropped from aircraft to detect submarines and may be encountered anywhere at sea. Other countries have similar sonobouys but their colour and dimensions are not known.

The above may frequently be met with in areas where H.M. Ships and Aircraft exercise, whether or not submarines are present, and should not be confused with submarine indicator buoys. In case of doubt the object should be approached to confirm, visually, whether or not it is a submarine indicator buoy before reporting it.

AIRCRAFT CASUALTIES AT SEA

Distress Communications

Visual Signals. An aircraft may indicate it is in distress by firing a succession of red pyrotechnic lights, by signalling "S.O.S." with signalling apparatus or by firing a parachute flare showing a red light. Navigation markers dropped by aircraft at sea, emitting smoke, or flames and smoke, should not be mistaken for distress signals. Low flying is not in itself an indication of distress.

An aircraft which has located another aircraft in distress may notify ships in the vicinity by passing a message in plain language by signalling lamp using the prefix "XXX." It may also give the following signals, together or separately, to attract a ship's attention:—*(a)* a succession of white pyrotechnic lights, *(b)* the repeated switching on and off of the aircraft's landing lights and *(c)* the irregular repeated switching on and off of the aircraft's navigational lights. If it wishes to guide a ship to the casualty or survivors it will fly low round the ship or cross the projected course of the ship close ahead at a low altitude opening and closing the throttle or changing the propeller pitch. It will then fly off in the direction in which the ship is to be led. British pilots are instructed to rock their aircraft laterally when flying off in the direction of the casualty. The ship should acknowledge receipt of the signals by hoisting the Code and Answering Pennant Close-up, or by flashing a succession of "T's" on the signal lamp, and may indicate the inbability to comply by hoisting the International Code Flag "N" or flashing a succession of "n's" on the signal lamp. it should then either follow the aircraft or indicate by visual or radio means that it is unable to comply. The procedure for cancelling these instructions is for the aircraft to cross the wake of the surface craft close astern at a low altitude, rocking the wings or opening and closing the throttle or changing the propeller pitch.

In order to take advantage of the greater visibility of pyrotechnics by night, searching aircraft will fly a creeping-line-ahead type of search, firing off green pyrotechnics at 5-10 minute intervals and watching for a replying red from the survivor.

Survivors from crashed aircraft in rubber liferafts may give the following distress signals:—

(1) Fire pyrotechnic signals emitting one or more red stars, or orange/red smoke.
(2) Flash a heliograph.
(3) Flash SOS or other distinctive signal by hand torch or other signalling lamp. Some liferafts may show a steady or a flashing light.
(4) Blow whistles.
(5) Use a fluorescein dye marker giving an extensive bright green colour to the sea around the survivors.
(6) Fly a yellow kite from the liferaft to support the aerial for the emergency radio transmitter.

Radio Signals. Radio is not carried by all civil aircraft. If, however, an aircraft transmits a distress message by radio, the first transmission is made on the designated air/ground route frequency in use at the time between the aircraft and the appropriate ground station, normally an Air Traffic Control Centre (A.T.C.C.). The aircraft might be asked by the A.T.C.C. to change to another frequency, possibly on another H.F. route frequency or on the civil aeronautical emergency frequency on 121·5 MHz in the VHF band. If the aircraft is unable to contact the ground station on the route frequency, any other available frequency may be used in an effort to establish contact with any land, mobile or direction-finder station. In addition, if time permits and the aircraft is so equipped, the distress call is made on the international distress frequency 500 kHz.

There is a close liaison among shore stations, including Air Traffic Control Centres, Rescue Co-ordination Centres, Coast Radio Stations and H.M. Coastguard, and merchant ships will ordinarily be informed of aircraft casualties at sea by broadcast messages from the Coast Radio Stations made on the international distress frequencies of 500 kHz, 2,182 kHz and VHF Ch. 16. Ships may, however, become aware of the casualty by:—
  (1) picking up an SOS message from an aircraft in distress which is able to transmit on 500 kHz or by intercepting a distress signal from an aircraft using radiotelephony on 2,182 kHz or VHF Ch. 16. (The form of such messages is given in Appendix A), or
  (2) by picking up a message from a search and rescue aircraft.

LIGHT-VESSEL—SIGNALS WHEN OUT OF POSITION

When any manned light-vessel off the coasts of the United Kingdom and Ireland is driven from her position to one where she is of no'use as a guide to shipping, the characteristic light will not be shown and the fog-signal will not be sounded, but the following signals will be made:—

By Day—Two large black balls will be shown, one forward and oine aft, and the International Code Signal L.O. indicating 'I am not in my correct position' will be hoisted.

By Day—A *red* fixed light will be shown at each end of the vessel and *red* and *white* flares shown simultaneously at least every 15 minutes; if the use of flares is impracticable, a *red* light and a *white* light will be displayed simultaneously for about a minute.

COLLISIONS WITH LIGHT-VESSELS

Caution—Light-vessels have been run into on several occasions by vessels navigating in their vicinity, and the lives of the men on board the light-vessel have been seriously endangered. The attention of mariners is drawn to the importance of making due allowance for the set of the tide and of exercising every precaution in order to GIVE ALL LIGHT-VESSELS A WIDE BERTH, especially when crossing their bows in a tideway, which should never be attempted unless absolutely necessary.

Under Section 666 of the Merchant Shipping Act, 1894, any person wilfully or negligently running foul of any light-ship or buoy is liable, in addition to the expense of making good any damage so occasioned, to a fine not exceeding £500 for each offence.

VESSELS NAVIGATED STERN FOREMOST

It has been agreed with the owners of British vessels chiefly concerned that the following signal should be displayed by vessels which are fitted with bow-rudders and are navigated stern foremost when entering or leaving certain ports and harbours in the United Kingdom and abroad, to indicate that for the time being they are navigating stern foremost:—

Two balls, each 2 feet in diameter, carried at the ends of a horizontal jackyard on the mast or, if the vessel has more than one mast, on the main or after-mast. The jackyard will be placed in a thwartship direction, at least 6 feet higher than the funnel top and will project at least 4 feet on either side of the mast so that the distance between the centres of the two balls will be at least 8 feet.

Bye-laws giving effect to this arrangement have been made for the ports of Dover, Ramsgate, Holyhead, Larne and Belfast.

BUOYS AND BEACONS

Wrecks have occurred through undue reliance on buoys and floating beacons always being maintained in their exact position.

They should be regarded simply as aids to navigation and not as infallible marks, especially when placed in exposed positions.

The lights shown by gas buoys cannot be implicitly relied on as, if occulting, the apparatus may get out of order, or the light may be altogether extinguished.

A ship should always, when possible, be navigated by bearings for angles of fixed object on shore and not by buoys or floating beacons.

VISUAL AND SOUND SIGNALS OF DISTRESS

Experience has shown that two of the existing statutory distress signals, viz. 'a continuous sounding with any fog-signal apparatus' and 'flames on the vessel' are not only liable to abuse, but when used as distress signals have often given rise to misunderstanding. A succession of signals on the whistle or siren is frequently made for other purposes than of indicating distress, e.g. for summoning a pilot, and may be mistaken for a 'continuous sounding'. similarly working lights and 'flare up' lights are authorised for use by fishing vessels and other small craft, and the simplest way of making a 'flare-up' light is to dip a rag in paraffin and set it alight. Unfortunately small vessels in distress frequently make the signal 'flames on the vessel' in the same manner. Thus it is often impossible to decide whether 'flare up' lights are being shown or whether distress signals are being made, especially in areas where fishing is carried on. As a result uncertainty and delay have occurred and lives have been lost in consequence.

Distress signals should be as distinctive as possible, so that they may be recognised at once and assistance despatched without delay. Thus, instead of making an indefinite succession of blasts on the fog-signalling apparatus when in distress, mariners should make the 'continuous sounding' by repeating the Morse Code signal SOS (· · · — — — · · ·) on the whistle or fog-horn. If this is done there can be no mistake as to the meaning of the signal. Similarly, by night, if signalling for help by means of a lamp or flashing light, the same signal, SOS, should always be used.

As regards the 'flames on the vessel signal', unless the flames making the signal are sufficiently large to attract immediate attention, the chances of being recognised as a distress signal are very poor. The best distress signal are *red* parachute flares or rockets emitting *red* stars, and whenever possible a supply of such signals should be carried. Arrangements should be made to steady rockets to ensure their satisfactory flight when fired. When it is not practicable to use the foregoing types of signals, lifeboat 5-star *red* signals, which can be held in the hand while being discharged, should be provided.

ROCKET SIGNALS, DISTRESS ROCKETS AND LINE-THROWING ROCKETS

Rocket signals, distress rockets and line-throwing rockets are liable to deteriorate if kept for a long period, and the Ministry of Transport have decided that they should be condemned after a period of two years from the date of manufacture.

Special care should be taken regarding the disposal of these obsolete fireworks. On no account should they be used for testing or practice purposes, or landed for any purpose. They should be kept in a safe place until opportunity occurs for throwing them overboard in deep water well away from land.

DAMAGE TO DRIFT NETS

Extensive damage has been caused to the nets of drifters fishing in the Firth of Forth, Irish Channel and off the South-West Coast of England and the Scilly Isles, through the failure of both cargo and passenger steamers to comply with the necessary precautions as laid down in the various Sailing Directions.

When passing through a fleet of drift net fishing vessels, it should be borne in mind that these nets often extend for a distance of 4 miles from the drifter, and every care should be taken to avoid damage to the nets when passing through the fleet.

In the Irish Channel where the nets are set from sundown to sunrise, masters are requested to exercise all reasonable care when in this position so as to avoid crossing the nets. Flares are displayed by fishermen when vessels are observed to be bearing down on their train of nets.

TANKERS—USE OF ROCKET LINE-THROWING APPARATUS

Attention is called to the danger of attempting to establish communication, by means of a rocket line-throwing apparatus, with an oil tanker, should that vessel be carrying petrol spirit or other highly inflammable liquid and be leaking. In such a case THE ASSISTING VESSEL SHOULD LIE TO WINDWARD OF THE TANKER and the communication should be established from the ship requiring assistance. THEREFORE BEFORE FIRING A ROCKET TO SUCH A VESSEL, IT SHOULD BE ASCERTAINED WHETHER IT IS SAFE TO DO SO.

When a vessel in distress is carrying petrol spirit or other highly inflammable liquid and is leaking, the following signals should be exhibited to show that it is dangerous to fire a line-carrying rocket by reason of the risk of ignition.

By Day.—Flag B of the International Code of Signals hoisted at the masthead.

By Night—A red light hoisted at the masthead.

When visibility is bad the above signals should be supplemented by the use of the following International Code signals made in sound—GU (— — — · · · —) 'It is not safe to fire a rocket'.

INFORMATION *RE* FOG-SIGNALS

The following information in regard to fog-signals is promulgated for the guidance of mariners:

(1) Fog-signals are heard at greatly varying distances.

(2) Under certain conditions of atmosphere, when an air fog-signal is a combination of high and low tones, one of the notes may be inaudible.

(3) There are occasionally areas around a fog-signal in which it is wholly inaudible.

(4) A fog may exist a short distance from a station and not be observed from it, so that the signal may not be sounded.

(5) Some fog-signals cannot be started at a moment's notice after signs of fog have been observed.

Mariners are therefore warned that fog-signals cannot be implicitly relied upon, and that *the practice of sounding should never be neglected.* Particular attention should be given to placing 'Look-out men' in positions in which the noises in the ship are least likely to interfere with the hearing of the sound of an air fog-signal; as experience shows that, though such a signal may not be heard from the deck or bridge when the engines are moving it may be heard when the ship is stopped, or from a quiet position. It may sometimes be heard from aloft, though not on deck.

There are three means adopted for signalling in fog:—

(a) By air sound signals comprising (1) *Diaphone,* (2) *Siren,* (3) *Reed,* (4) *Nautophone,* (5) *Gun,* (6) *Explosive,* (7) *Bell* or *Gong,* and (8) *Whistle.*

(b) By submarine sound signal produced either by (9) an *Oscillator* or (10) *Bell,* and

(c) By Wireless Telegraphy.

I. Air Fog-Signals

The *Diaphone* (1), *Siren* (2), and *Reed* (3) are all three compressed air instruments fitted with horns for distributing the sound.

The *Diaphone* emits a powerful low-tone note terminating with sharp descending note termed the 'grunt'; the *Siren* a medium-powered note, either high or low or a combination of the two, and the *Reed,* a high note of less power. *Reeds* may be hand operated, in which case the signals from them are of small power.

The *Nautophone* (4) is an electrically operated instrument also fitted with a horn, and emits a high note signal similar in power and tone to that of the Reed.

Gun (5) and *Explosive* (6) signals are produced by firing of explosive charges, the former being discharged from a gun and the latter being exploded in mid-air.

Bells (7) may be operated either mechanically or by wave action, in which latter case the sound is irregular. The notes may be high, medium or low according to the weight of the bell. *Gongs* are also sometimes employed.

A *Whistle* (8) is a signal of low power and tone sometimes fitted on a floating body; when this is the case the sound is produced by air drawn in and compressed during the upward and downward movement of the body due to wave action, and is consequently irregular.

II. Submarine Sound Signals

The *Oscillator* (9) is an electrically operated instrument sounding a high note signal.

Bells (10) may be operated either mechanically or by wave motion, in which latter case the sound is irregular.